CROP SPRAYING AND THE HEALTH OF RESIDENTS AND BYSTANDERS

CROP SPRAYING AND THE HEALTH OF RESIDENTS AND BYSTANDERS

THE ROYAL COMMISSION ON ENVIRONMENTAL POLLUTION

Nova Science
New York

For permission to use material from this book please contact us:
Telephone 631-231-7269; Fax 631-231-8175
Web Site: http://www. novapublishers. com

NOTICE TO THE READER

Library of Congress Cataloging-in-Publication Data:
Crop spraying and the health of residents and bystanders / Tom Blundell, editor.
p. cm.
ISBN 1-60021-391-X
1. Pesticides--Health aspects--Great Britain. 2. Pesticides--Environmental aspects--Great Britain. 3. Pesticides--Government policy--Great Britain. I. Blundell, Tom, Sir.
RA1270.P4C76 2006
363.738'498--dc22 2006024514

Published by Nova Science Publishers, Inc. ✦ New York

CONTENTS

Preface

This report addresses a complex and controversial issue: the human health risks associated with the use of agricultural pesticides. Some people are extremely concerned about their potential exposure to pesticides, arising because they occupy properties adjacent to farmland or because they have (or have had) access to such land, for example when using footpaths.

The official position in the UK in response to these concerns has been that a robust approach to the assessment of human health risks associated with pesticide exposure already exists, and that there is no scientific case for taking additional measures, such as the introduction of no-spray buffer zones, to protect members of the public who may be in the vicinity of a sprayed area. But those who consider themselves to be adversely affected by pesticide spraying have not been reassured, and have continued to campaign for the adoption of more precautionary measures.

About the Royal Commission on Environmental Pollution*

The Royal Commission on Environmental Pollution is an independent standing body established in 1970 to provide authoritative advice on environmental issues. Its terms of reference are:

> To advise on matters, both national and international, concerning the pollution of the environment; on the adequacy of research in this field; and the future possibilities of danger to the environment.

Within this remit the Commission is free to consider and advise on any matter it chooses; the UK government or the devolved administrations may also ask it to consider particular topics.

The primary function of the Commission is to contribute to policy development in the longer term by providing a factual basis for policy-making and debate, and setting new agendas and priorities. It considers the economic, ethical and social aspect of issues alongside the scientific and technological aspects. It sees its role as reviewing and anticipating trends and developments, identifying fields where insufficient attention is being given to environmental problems, and recommending actions that should be taken. The Commission has published 25 reports, and many of their recommendations have been accepted and implemented by successive governments.

The Members of the Commission have a wide range of expertise and experience in natural and social sciences, medicine, engineering, law, economics, and business. They serve part-time and as individuals, not as representatives of organisations or professions.

A full-time Secretariat supports the Chairman and Members by arranging and recording meetings and visits; gathering and analysing information; handling finances and administration; and drafting and publishing the Commission's reports.

In the course of its studies, the Commission canvasses a wide range of views. Information on its work (including minutes of meetings, background papers by consultants and summaries of evidence submitted) is available via **www. rcep. org. uk.**

A special report produced on request of the Department for Environment, Food and Rural Affairs by the ROYAL COMMISSION ON ENVIRONMENTAL POLLUTION

---

* Extracted from Royal Commission on Environmental Pollution

In: Crop Spraying and the Health of Residents... ISBN: 1-60021-391-X
Editor: The Royal Commission, pp. 1-15

Chapter 1

INTRODUCTION

1.1. This report addresses a complex and controversial issue: the human health risks associated with the use of agricultural pesticides. Some people are extremely concerned about their potential exposure to pesticides, arising because they occupy properties adjacent to farmland or because they have (or have had) access to such land, for example when using footpaths.

1.2. The official position in the UK in response to these concerns has been that a robust approach to the assessment of human health risks associated with pesticide exposure already exists; and that there is no scientific case for taking additional measures, such as the introduction of no-spray buffer zones, to protect members of the public who may be in the vicinity of a sprayed area. But those who consider themselves to be adversely affected by pesticide spraying have not been reassured, and have continued to campaign for the adoption of more precautionary measures.

REQUEST FROM THE MINISTER

1.3. In June 2004 the Rt Hon Alun Michael, then Minister for Rural Affairs and Local Environmental Quality, asked the Royal Commission on Environmental Pollution to undertake 'a study into the science used to assess risk to people from crop spraying'. [1] The Minister asked for the study as a 'key part of Defra's [the Department for Environment, Food and Rural Affairs] response to two public consultations' launched in 2003, [2] the first to obtain views on the introduction of no-spray buffer zones between fields and residential properties, and the second to invite suggestions on measures to improve public access to information on pesticide use. These consultations were originally launched following the claims about health impacts noted above.

1.4. According to Defra there were a total of 763 responses to the consultation on buffer zones and 487 responses to the consultation on access to information. Responses were polarised; members of the public and environmental groups were largely in favour of buffer zones and access to information, whereas most farmers were against these proposals. Approximately equal numbers of those who responded to the consultation on buffer zones, were for and against the proposals. For the consultation on access to information on pesticide use, 375 were in favour of increased access, and 109 were opposed. [3]

1.5. Following these consultations the Minister said 'the Government's top priority is to ensure that the safety arrangements we have in place protect the public. The independent scientific advice to me is very clear that the existing system provides full reassurance on that score. For this reason I have decided against the introduction of compulsory no-spray "buffer zones"around agricultural land. '[4]

1.6. On the subject of access to information the Minister announced:

- new legal measures requiring farmers and growers to keep records of pesticides sprayed on crops and to make those records available to the public via a third party;
- a pilot study to explore how residents living next to farms can be notified ahead of spraying; and
- the Minister concluded '... despite existing advice, there is clearly a perception that current arrangements are inadequate. I have listened to the concerns of campaigners who hold strong views about how crop spraying has affected their health. I believe the time is now right for a fresh and independent appraisal of the basis for risk assessment. That is why I have asked the Royal Commission on Environmental Pollution to examine the evidence on which the current system is based and the reasons for people's concerns'. [5] (See appendix A for the Ministerial statement).

1.7. In response to this request, the Commission announced its study to examine the scientific evidence on which Defra based its decision on buffer zones and its policy on access to information on crop spraying. To this we added the consideration of wider issues related to the handling and communication of risk and uncertainty, as well as public involvement, values and perceptions in the context of agricultural pesticide use. [6] (See appendix B for the Commission's announcement of study and invitation to submit evidence).

Pesticides and Health

1.8. The UK Food and Environment Protection Act (FEPA) 1985 defines a pesticide as 'any substance, preparation or organism prepared or used, among other uses, to protect plants or wood or other plant products from harmful organisms; to regulate the growth of plants; to give protection against harmful creatures; or to render such creatures harmless'. [7] Pesticides are widely used in food and commodity production and storage, the promotion of public health, and in the preservation and maintenance of buildings and structures. [8] The term pesticide therefore embraces a diverse group of chemicals (herbicides, insecticides, fungicides, rodenticides and soil sterilants among others), which vary widely in their mode of action and their potential impact on organisms (table E. 1 in appendix E). In general, however, all pesticides work by interacting with target pest organisms, usually by disrupting an essential biological process, for example by affecting nerve transmission or mimicking hormones.

1.9. The term pesticide is used in two ways. It can mean either the active ingredient that has the pesticidal property or the final formulation, which includes a mixture of chemicals (co-formulants) that aim to improve the function of the pesticide. According to the Pesticides Safety Directorate (PSD), the UK agency responsible for agricultural pesticide policy and approval, currently around 275 active ingredients are authorised for use as pesticides in the UK, and around 1, 000 different co-formulants can be used in the final formulations. The approval of pesticides is in transition from a national system based on FEPA 1985 and the Control of Pesticides Regulations 1986 (as amended) to a European system set out in Council Directive 91/414/EEC on plant protection products (5. 2-5. 6). A full list of pesticides registered for use in agriculture, horticulture, forestry and amenity applications in the UK can be found on the PSD website. [9]

1.10. The hazardous properties of chemicals are generally considered to occur in relation to their persistence, bioaccumulation and toxicity (PBT) properties. Persistence is a measure of how long a substance lasts in the environment and is usually expressed as the half-life of the substance, which is the time taken for degradation of 50% of the initial concentration of a substance in a particular medium such as soil or water.

Bioaccumulation is the degree to which a substance accumulates in various (normally fatty) tissues of living organisms. In the absence of a measured or experimental bioaccumulation or bioconcentration factor this is estimated using the octanol-water partition coefficient to indicate the fat solubility.

1.11. Pesticides are designed to be biologically active, and so may interact with a wide range of biological systems. However pesticides are designed to be selectively toxic against target pest organisms as opposed to humans and non-target organisms. New pesticides usually have a more selective toxicity than those they replace.

1.12. It is unlikely that we will ever be able to say with complete certainty that a pesticide has no harmful effects on non-target organisms. The aim of pesticides policy is to try to ensure that people and non-target organisms are not exposed to amounts above those at which tests on animals, or pre-existing data, suggest there could be deleterious effects.

In practice, the regulators aim to reduce exposure to levels lower than this with the use of uncertainty factors in the approval of pesticide active ingredients (2. 49-2. 50), and through restrictions and guidance on pesticide use. The risk assessment process undertaken for pesticides is discussed in further detail in 2. 40-2. 50.

1.13. All pesticides have to be tested to demonstrate their highly selective and effective action against the target organism and to determine their potential toxicity to other organisms, including humans (table E. 1 in appendix E and 2. 43-2. 48). The government's stated objective for toxicological testing on active ingredients is that it should be comprehensive and undertaken for a well-defined range of health effects. But for the final product (active ingredient mixed with co-formulants) only limited tests are undertaken to look for acute and local effects. Toxicity testing currently provides little information about the chronic multisystem health effects that are attributed to pesticide exposure in humans (2. 10-2. 15). Many of these effects are, in any case, poorly understood. Testing for chronic effects requires repeated dose tests carried out over long periods in the life of the test organisms. Currently there are few animal models, or other viable means, to test for all of these chronic and multisystem illnesses.

1.14. Pesticide handling and use procedures are designed to minimise non-target exposure to levels below those at which acute and chronic effects are predicted. Ideally effective systems for surveillance and biological and epidemiological monitoring should be in place to gather data to provide a check on the risk assessment that was undertaken by the regulators prior to the pesticide's approval and provide evidence to assess whether or not there have been unforeseen effects on health from any exposure that may have taken place.

1.15. Effective monitoring for unforeseen health effects is difficult except for well-defined groups such as spray operators. We did not undertake our own comprehensive critical review of the health-based literature for either occupational or non-occupational exposure or use health-based evidence in an attempt to attribute causality in either group but there is significant literature on possible linkages between exposure to pesticides and a variety of health effects in humans. Some studies indicate a statistically significant correlation and some do not. Several well-publicised reports have recently attempted to analyse the epidemiological literature concerning pesticide exposure and adverse health effects (2. 20-2. 34).

1.16. Although some people have reported acute health effects from a single exposure, most of the concerns reported to us relate to long-term (chronic) ill health.

The situation is complex. If there is an effect, it could relate to a single acute exposure with a long latency or to repeated doses over time, that can be described as chronic exposure. Linking chronic health effects to a particular exposure or pesticide will not usually be possible. Different pesticides are used during the year to deal with different problems (e. g. herbicide before the crop establishes, then insecticides and fungicides while it is growing), or from year to year as different crops are grown. The difficulty is further compounded by changes in the pesticides available commercially. New pesticides are developed and older ones withdrawn. Some people may be more vulnerable than others for reasons such as existing chronic health conditions, genetic make-up and age, including those still in the womb. The exposure—response relationship for a given pesticide is likely to vary between individuals. This is a more realistic position than the assumption that there is a single safe dose for everyone. [10] This is reason why uncertainty factors are widely employed once the key toxicological parameters have been established.

Resident and Bystander Exposure to Agricultural Pesticide Spray

1.17. The primary routes of human exposure to pesticides are via pesticide use (occupational and in the home and garden), consumption of food and water, living near to where pesticides are applied and bystander exposure. This report focuses on issues related to resident and bystander exposure in the context of agricultural pesticides. We use the term agriculture to include horticulture and commercial forestry. In the course of our consultations we have also heard from members of the public who have attributed health problems to timber treatment and amenity spraying (for example, on highways, parks and golf courses), but in the limited time available we have not been able to extend the scope of our study to include these issues.

1.18. The spraying of crops can result in pesticide drift away from the target:fine droplets and particles carried by air currents until they contact a surface, evaporate or degrade.

The term 'spray drift' is commonly employed to mean the movement of a pesticide droplet 'through the air at the time of application or soon thereafter, to any site other than that intended for application (often referred to as off-target)'. [11] Crop spraying may also result in the movement of pesticides to off-target sites caused by erosion, migration, volatility, or contaminated soil particles that are windblown after application. In this study we are concerned with the total effect of crop spray and in chapter 3 we discuss vapour transfer explicitly, and consider this and particle movement as potential routes of exposure. The contamination of water or marketed food from crop spray is outside the remit of this report.

Residents and Bystanders

1.19. We adopt the term 'bystander'in recognition of its currency in debates about human exposure to pesticides. It is not a term we would have chosen. It is seen by some of those affected as failing to recognise the permanency of their residence in areas close to where crops are sprayed, the degree of their exposure, or even the legitimacy of their role in rural society. Few people who may be affected by agricultural crop spray could be regarded as bystanders as the term is used in common parlance, to mean a non-participating spectator. We therefore use the term 'residents' for those who live near to sprayed fields. The latter includes:

- those working, at school or in hospital in the vicinity of fields that are sprayed;
- those walking, cycling or riding on public highways, bridleways or rights of way on private land; and
- legitimate visitors and trespassers on land that is either directly sprayed or subject to spray drift.

1.20. Where the term bystander appears in this report in quotation marks, this should be read in a wider sense to include residents as that is the way in which most of the outside bodies view the term.

1.21. Many villages in the United Kingdom are surrounded by fields that are sprayed with pesticides (figure 1-I), with many properties and gardens in rural areas adjacent to such fields (figure 1-II).

1.22. Research into the number of residents and bystanders in the UK who may be exposed to pesticides as a result of crop spraying has not been undertaken. Pesticide Action Network (PAN) *UK*, a pressure group, provided a preliminary estimate of 250, 000 (for England and Wales) in 2003. [12] PAN *UK* acknowledges that this figure is an extrapolation only and has urged Defra to commission research to obtain more accurate data.

In the absence of such data we commissioned the Centre for Ecology and Hydrology (CEH) to calculate a figure. CEH estimated the total length of boundaries between residential and arable and horticultural land in Great Britain at about 20, 700 km; the number of properties that fall within land immediately bordering arable and horticultural land at approximately half a million; and the total number of people resident in these properties at between 1 million and 1. 5 million (based on 1998 data). The full CEH report, including details of the methodology employed and limitations thereof, can be found on the Commission website. [13]

Figure 1-I. Aerial photo showing rural settlements surrounded by arable land[14].

Figure 1-II. Proximity of residential property to arable land[15].

1.23. The length of footpaths, tracks, roads and canals adjacent to (within 6 metres) or running through arable and horticultural land is estimated at 253, 000 km. [16] It seems reasonable to assume that many of these routes are used on a regular basis, though the numbers of people using them soon after pesticide spraying have not been estimated.

1.24. We do not know the actual numbers of residents and bystanders in the UK who claim they are suffering ill health caused by exposure to pesticides as a result of crop spraying.

PAN *UK* has 973 entries on its Action on Pesticide Exposure (PEX) database, which are classified as people who have been exposed to any pesticide, the majority of which are residents or bystanders. In written evidence, rural resident and pesticides campaigner, Georgina Downs told us that in her database there are entries for 464 people reporting ill health, which they associate with pesticides, and that over three-quarters of these reports are specifically related by claimants to crop spraying. These figures are related to the period early 2001 to June 2005. About 30 of these cases relate to countries other than the UK. [17] In the Health and Safety Executive (HSE) *Pesticide Incidents Report*, it is stated that the Pesticide Incidents Appraisal Panel (PIAP) (2. 80) received information on 59 incidents of alleged ill health in members of the public from all pesticides in 2003/4, of which only one case was classified as a confirmed case and 14 as likely cases. [18] In the course of this study we have received over 50 submissions of evidence of ill health alleged to be the result of agricultural pesticide exposure. We have also has access to the submissions to the PSD consultations (2. 7). Although there is considerable uncertainty about the number of people who are ill, it does appear to be a very small proportion of those potentially exposed.

Agricultural Spraying: Current Practice

1.25. Of the 18. 4 million hectares (ha) of agricultural land in the UK, 4. 6 million ha is farmed for arable and horticultural crops. Although some spraying takes place on other land, it is predominately arable land that is intensively sprayed and it is only in relation to arable land and horticultural cropping that we have heard concerns from the public.

Of the 4. 6 million ha only 24, 000 ha (0. 5%) are organically farmed. The average number of times a field is sprayed per year is five (3. 4). [19] Table 1. 1 shows the total weight (in kg) of pesticide active ingredients applied in recent years to all crops (horticultural, arable, grassland and fodder crops) in the UK. The amount of active ingredient used has decreased by approximately 14% in the period 1998-2003. An earlier reduction was said to be due largely to a continued move to newer, more active molecules applied at lower rates, together with a more widespread adoption of reduced rate applications of pesticides, particularly for fungicides where disease pressure is low and herbicides where several reduced rate applications may often still result in a lower total amount of pesticide applied than a single full rate application. [20] Similar factors are likely to underlie this more recent reduction.

Table 1. 1. Total weight (in kg) of pesticide active ingredients applied in recent years to all crops in the UK[21]

YEAR	TOTAL WEIGHT ACTIVE INGREDIENT APPLIED (KG)
1998	35, 485, 618
1999	35, 238, 494
2000	32, 906, 101
2001	32, 766, 438
2002	30, 830, 175
2003	30, 545, 618

1.26. An estimated 49, 000 farmers and employees[22] and approximately 200 professional agricultural contract spraying businesses[23] are involved in spraying on arable farms.

There are 1, 334 contractor sprayer operators listed on the National Register of Sprayer Operators (NRoSO). [24] In 2004 18% of the UK arable area was treated exclusively by contractors, whilst holdings accounting for 72% of the UK arable area did not use a contractor at all. A combination of both contractors and farm staff, with the latter making most applications, was used on the remaining 10% of land. The total area sprayed by contractors accounted for 20% of the total UK arable area. [25]

1.27. Spraying is done primarily with the use of boom sprayers for arable crops and field vegetables (figure 3-I), and fan-assisted sprayers for orchards (figure 3-II). The use of all terrain vehicles, such as quad bikes, predominates in upland areas, which are generally not arable. Smaller scale applications are made with hand-held sprayers. Some pesticides are applied by injection into the soil or by painting onto plant surfaces (for example, for contact herbicides). Crops can also be sprayed by air, although we understand that in the UK this practice has all but disappeared.

1.28. Crops can be sprayed up to the boundary of a field and can be close to neighbouring property, residents or bystanders (figure 1-III). Spray drift could lead to the direct exposure of residents and bystanders through contact with the skin or by inhalation.

It could also lead to deposition on property or public rights of way leaving pesticide residues, through which people might be exposed (for example, by contact with contaminated surfaces or ingestion of contaminated food grown in gardens).

Volatilisation of the pesticide from plants or soil could lead to subsequent atmospheric transport into neighbouring property and low-level but longer lasting exposure of residents through inhalation.

1.29. For approximately 135 active ingredients (or combinations of active ingredients)[27] used in the UK there is a statutory requirement to have five metre or one metre[1] strips of land alongside watercourses, known as 'buffer zones', or 'buffer strips', in which these pesticides are not applied because they are highly toxic to aquatic life. Pesticides can also contaminate water used to supply drinking water. The width of a buffer zone may be reduced for certain products[2] by applying a Local Environment Risk Assessment for

[1] Five metre buffer zones are required when ground crop sprayers are used whilst a one metre zone is required for hand-held sprayers.

[2] These are known as category B products, a list of which can be found at: http://www. pesticides. gov. uk/PSD_Databases/products/catAB-ap. cfm

Pesticides (LERAP). A reduced buffer zone usually involves the use of special spray nozzles which reduce drift (3. 11). A number of schemes (5. 101-5. 109 and 5. 113-5. 115) encourage the use of buffer zones. In response to public requests, or simply as good practice, some farmers have voluntarily introduced buffer zones in situations where exposure of residents and bystanders could occur (figure 1-III).

Figure 1-III. Some people live close to sprayed fields, however, the voluntary use of buffer zones can increase the distance from crop spraying, as in this picture[26].

Some spray crops at times of the day when potential exposure is thought to be minimised (in the early morning or late evening).

1.30. There is a requirement to notify and seek the agreement of the Environment Agency when spraying certain herbicides near watercourses. It is also a statutory requirement to tell residents and bystanders before spraying sulphuric acid, or before the aerial spraying of pesticides, but this does not apply to other pesticides or application methods. In the case of sulphuric acid, which is used to dehydrate potato haulms, notices must be posted on adjacent roads and paths warning passers-by and drivers of vehicles of the time and place of the intended application and of the possibility of hazard. These notices must be kept in place for at least 96 hours. In addition, at least 24 hours written notice of the intended operation and the possibility of a hazard must be given to occupants of any premises and to the owner, or his agent, of any livestock or crops within 25 metres of any boundary of the land intended to be treated. These conditions are outlined in the approval schedule for sulphuric acid. [28]

1.31. Schedule 4 to the Control of Pesticides Regulations 1986 (as amended) prohibits the aerial application of a pesticide unless the spray operator, or someone designated in writing on their behalf, has, amongst other things, given notice of the intended application to the occupants or their agents of all property within 25 metres of the boundary of the land to which the pesticide is to be applied. Notice must be given not less than 24 hours and (so far as is reasonably practicable) not more than 48 hours before the commencement of the aerial application. [29]

1.32. However, these requirements for notifying the public only apply to certain types of pesticides and methods of application. We are told that the pilot project for prior notification of all pesticide application announced by Minister Alun Michael in June 2004 (1. 6), is underway and the conclusions will go to Ministers in late summer 2005.

1.33. The Minister also announced new legal measures to oblige farmers and growers to keep records and provide information, via a third party, about which pesticides are being sprayed (1. 6). In the meantime, EC Regulation 852/2004[30] introduced an overriding statutory obligation for food business operators producing or harvesting plant products to keep records on:

- any use of plant protection products and biocides;
- any occurrence of pests or diseases that may affect the safety of products of plant origin; and
- the results of any relevant analyses carried out on samples taken from plants or other samples that have importance to human health.

1.34. This comes into effect on 1 January 2006. The government is still considering the procedures that will be introduced to make such records publicly available (5. 78-5. 84).

1.35. The exposure of residents and bystanders to pesticides depends on a wide range of factors and operating conditions including the pesticide used, wind speed and direction, and boom height. This means that the amount of pesticide to which a resident or bystander may be exposed will vary over a wide range from low to potentially quite high levels in the worst case. Given that both the level of exposure and the level of susceptibility to that exposure may vary between individuals (2. 23), ill health might arise if a person who is significantly more susceptible than average was exposed to a higher than average amount of pesticide. One explanation of the small number of reported cases of ill health from the exposed population could be the rare occurrence of this combination of circumstances. We examine this in more detail in chapters 2 and 3.

Bystander Exposure: Concern and Activity in Other Countries

1.36. This report focuses on resident and bystander exposure primarily in the United Kingdom but takes account of experiences elsewhere in the world. We have received evidence on resident and bystander exposure from a number of countries and it is clear that the level of concern and activity surrounding the issue varies considerably between countries. Governments in the United States, parts of Europe, New Zealand and Canada appear to have shown some interest, many others not so.

1.37. In the United States regulations that specifically address the issue of resident and bystander exposure vary by state and no aspects of state regulation are universally applicable. [31] Arizona and New Jersey have restricted ground spraying around schools for certain pesticides. California has a central database recording agricultural spray application, which includes summaries of pesticide use by crop, product, chemical, location and date. [32] Recent press reports from Florida mention allegations from the ex-head of the State Health

Department pesticide monitoring programme that concerns about ill health following aerial spraying to control fruit fly had been suppressed. [33] A few states, for example Kansas and Washington, have in effect created drift regulations through judicial decision. There is no evidence yet as to whether any of these actions have reduced claims of ill health.

1.38. In Europe the picture is also varied. In Italy there are no specific regulations to control pesticide spray for residents and bystanders in a systematic way, but the issue is a cause of some concern in the light of complaints which the government receives from citizens living near sprayed crops. [34] Resident and bystander exposure to pesticides in Sweden does not seem to be of much public concern. This may be due to already strong regulation and general trust in the system to restrict use of pesticides. [35] In Denmark there are no laws to regulate the distance that pesticide spraying must be from a neighbour's land, but good farming practice recommends that a minimum distance should be upheld. The issue is not a major cause of concern, perhaps also because Denmark has one of the lowest pesticide usage rates in Europe. [36] In the Netherlands there is no procedure to evaluate the exposure of people living near pesticide spraying and no statutory base for buffer zones. [37] The French Department of Health has been running a pesticides and health programme since October 2001, which includes research into exposure to pesticides of the population that lives close to agricultural areas. [38]

1.39. In New Zealand some councils give households the option to register as a 'no spray' property with respect to council roadside weed control programmes. In return the householder must control the weeds themselves.

1.40. Crop spray and resident and bystander exposure are a matter for concern in Canada.

Pesticides sold or used in Canada are evaluated by the Environmental Health Directorate to determine the possible hazards to residents and bystanders. In most provinces pesticide applications on public land, and by pest control operators on residential property, require the posting of signs. [39] Largely in response to health concerns a number of Canadian municipalities, including the three largest and the province of Quebec, have enacted bans on the cosmetic use of pesticides – non-essential use where the application is purely for an aesthetic pursuit. [40] This is mostly on lawns and in gardens. Quebec further prohibited the use of pesticides inside and near to day-care centres, schools and summer camps, citing the heightened vulnerability of children. [41]

Evidence

1.41. The issue of resident and bystander exposure to agricultural pesticides involves a diverse cross-section of society. We use the term 'public'in this report to denote the plurality and diversity of opinions and interests and not to imply a homogeneous group.

1.42. Our focus in this report is on the issue of resident and bystander exposure, and specifically on the risks to health that may be posed by crop spraying. We are conscious, however, that divergent positions on 'wider'questions, such as current levels of pesticide use or the merits of alternative systems of agricultural production, may underlie controversies about risks to health and the adequacy of public protection.

In taking evidence it became clear to us that such issues were indeed significant, though in the context of this study we have not attempted to address them in any depth.

1.43. We called for written evidence[3] from a wide range of organisations and individuals and obtained copies of submissions to the two public consultations launched by Defra in 2003 into buffer zones and access to information. In addition we received oral evidence, [4] met a number of individuals and organisations, and made visits to residents (2. 6-2. 8), farms, general practitioners, the Silsoe Research Institute, the Institute for Environment and Health, the Small Area Health Statistics Unit at Imperial College of Science, Technology and Medicine, London, and the Syngenta Central Toxicology Laboratory (appendix C).

1.44. To engage a range of perspectives in framing the scope of the study and in defining the questions to be addressed, we held an open meeting at the start of the study, on 25 September 2004. [5] Speakers identified key issues to be considered in the study (appendix D) and an open forum provided opportunity for debate. Participants were invited to make written contributions in response to speakers'presentations and more generally on the subject of resident and bystander exposure to crop spray.

1.45. The open meeting identified a range of issues and contextual questions to be addressed, which informed the conduct of our study (table 1. 2). These were augmented by submissions, oral evidence and visits.

Table 1. 2 . Questions to be addressed in the study arising from the open meeting

	QUESTIONS
HEALTH	What are the biological effects of pesticides on humans? Can science currently explain the health effects that people are attributing to agricultural pesticide use? Are current systems able to pick up health effects that are not well defined in the scientific literature? Are general practitioners and clinicians responding effectively to health problems reported to them? How important is it to adopt a more precautionary approach to exposure?
MODELLIN G AND EXPOSURE	Is the scientific basis for assessing exposure to pesticides robust? Are all routes of bystander exposure to pesticides accounted for in the assessment? How reliable are the exposure estimates, and data on which they are based?
LEGAL	What level of protection is provided by the legal system in cases of bystander exposure? Is the law in need of reform?
POLICY	How have pesticide regulators handled bystander exposure? How should bystanders be informed about agricultural pesticide use and notified of spray activity? How effective is the Green Code and how well is it observed, monitored and enforced? Does the National Pesticide Strategy address public concerns about pesticides? How are standards for risk and uncertainty established?
OTHER CONCERNS	What is the scale of bystander exposure? What economic impact would measures to control exposure have?

1.46. We took a conscious decision to start by standing back from the detail, taking a fresh look at the nature of the concerns put to us and the strength of the evidence in each case

[3] This evidence is available in the Royal Commission on Environmental Pollution library.
[4] Transcripts of oral evidence sessions can be found on the Royal Commission on Environmental Pollution website.
[5] In line with procedures advocated in our Twenty-first Report, *Setting Environmental Standards*.

and in totality. We are grateful for the help we have received in this task from all those who have made submissions to us.

STRUCTURE OF THE REPORT

1.47. Chapter 2 recognises the complex nature of health concerns. It examines: effects attributed by residents and bystanders to pesticide exposure; the use of animal models for assessing acute and chronic health effects in humans; the potential of current epidemiology in cases of resident and bystander exposure to pesticides; and systems that are in place to treat and monitor the health problems ascribed to pesticide exposure.

1.48. Chapter 3 examines the factors determining resident and bystander exposure to crop spraying and the approach used by the UK government to assess resident and bystander exposure to pesticides. It considers the adequacy of that approach in the light of analyses undertaken by the UK regulators.

1.49. Chapter 4 examines the level of redress offered to residents and bystanders by private law remedies.

1.50. Chapter 5 examines the mechanisms of agricultural pesticide governance. In particular we consider the approval process, controls on pesticide use, and monitoring and enforcement of the regulations. These include looking at the policy framework, governmental organisations, regulations (including at the European level), and nongovernmental schemes.

1.51. Our investigations have led us to the conclusion that the examination and reporting of chronic health effects in residents and bystanders needs to be taken more seriously; that the exposure assessment needs further research and development; that information about pesticide spraying needs to be more freely available; that the governance of pesticides relating to approvals and scientific advice needs reform; that the rules governing the actual application of pesticides need to be tightened and better enforced; and that some precautionary measures should be introduced to give further protection to residents and bystanders, especially those who may be most susceptible to pesticide spraying. Our conclusions and recommendations are developed in the chapters that follow and summarised in chapter 6.

REFERENCES

[1] Department for Environment, Food and Rural Affairs (2004). *News Release:Royal Commission to Examine Science on Human Exposure Risks from Crop Spraying*, http://www. defra. gov. uk/news/2004/040616d. htm accessed July 2005.

[2] *Ibid.*

[3] *Ibid.*

[4] *Ibid.*

[5] *Ibid.*

[6] Royal Commission on Environmental Pollution (2004), *News Release:Royal Commission Agrees to Study of Pesticides and Bystander Exposure*, http://www. rcep. org. uk/news/04-06. htm accessed July 2005.

[7] *Food and Environment Protection Act 1985*. TSO 1985.

[8] Hurst, P. , Hay, A. and Dudley, N. (1991). *Pesticides Handbook*, Journeyman Press.

[9] Pesticides Safety Directorate, http://www. pesticides. gov. uk/home. asp accessed April 2005.

[10] Evidence from the Department of Health, March 2003.

[11] United States Environmental Protection Agency, *Spray Drift of Pesticides*, http://www. cdpr. ca. gov/docs/factshts/epadoc. htm accessed July 2005.

[12] Pesticide Action Network UK, *Consultation response on buffer no-spray zones from Pesticide Action Network UK*, submitted to the Pesticides Safety Directorate on October 2003.

[13] Centre for Ecology and Hydrology (2005) *National estimates of residential land adjacent to arable land using Countryside Survey sample data*, www. rcep. org. uk/pesticides.

[14] Supplied by Bluesky Data © (2005).

[15] Supplied by Philippa Powell (2005).

[16] Centre for Ecology and Hydrology (2005)

[17] Evidence from Ms Georgina Downs, June 2005.

[18] Health and Safety Executive (HSE) *Pesticide Incidents Report*, http://www. hse. gov. uk/fod/pir0304. pdf accessed May 2005.

[19] Central Science Laboratory, *Pesticides Usage Survey Report 187: Arable Crops in Great Britain 2002*, http://www. csl. gov. uk/science/organ/pvm/puskm/arable2002. pdf accessed April 2005.

[20] Central Science Laboratory, *Pesticides Usage Survey Report 150: Review of Usage of Pesticides in Agriculture and Horticulture throughout Great Britain 1986-1996*, http://www. csl. gov. uk/science/organ/pvm/puskm/report150. pdf, accessed May 2005.

[21] Central Science Laboratory, Pesticide Usage Statistics, http://pusstats. csl. gov. uk/myresults. cfm accessed May 2005.

[22] Central Science Laboratory, *Pesticides Survey Report:A Survey of Current Farm Sprayer Practices in the United Kingdom 2004*, http://www. voluntaryinitiative. org. uk/_Attachments/Second%20Farm%20Application%20Practice%20Survey% 20June%202005. pdf accessed July 2005.

[23] Evidence from the National Association for Agricultural Contractors (NAAC), October 2004.

[24] Evidence from the NAAC, April 2005.

[25] Evidence from Dr Miles R Thomas, Pesticide Usage Survey and Knowledge Management Systems, April 2005.

[26] Supplied by Charlotte Grey (2005).

[27] Pesticides Safety Directorate, *LERAPs (Buffer Zone): Search by Category A and B Active Ingredients*, http://www. pesticides. gov. uk/PSD_Databases/products/catAB-fp. cfm accessed April 2005

[28] Pesticides Safety Directorate, *Schedule: Commodity Substance: Sulphuric Acid*, http://www. pesticides. gov. uk/approvals. asp?id=311 accessed April 2005.

[29] *The Control of Pesticides (Amendment) Regulations 1997* (Schedule 4, Conditions relating to consent to the use of pesticides by aerial application), Statutory Instrument SI 1997 No. 188, http://www. legislation. hmso. gov. uk/si/si1997/97018802. htm accessed April 2005.

[30] EC Council Regulation COM 852/2004 of 30 April 2004 on the hygiene of foodstuffs, Official Journal of the European Communities, **L139**, 30. 04. 2004.

[31] Feitshans, T. A. (1999). An Analysis of State Pesticide Drift Laws, *San Joaquin Agricultural Law Review*, **37**

[32] Department of Pesticide Regulation, *California Pesticide Information Portal*, http://calpip. cdpr. ca. gov/cfdocs/calpip/prod/main. cfm accessed May 2005.

[33] Lanitgua, A. *Pesticide Watchdog called ag lap dog*, PalmBeachPost. com, http://www. palmbeachpost. com/localnews/content/local_news/epaper/2005/05/01/m1a_whistleblow-er_0501. html accessed May 2005.

[34] Evidence from the Foreign and Commonwealth Office, December 2004.

[35] *Ibid.*

[36] *Ibid.*

[37] *Ibid.*

[38] *Ibid.*

[39] Pest Management Regulatory Agency (2003). *Fact Sheet on The Regulation of Pesticides In Canada*, http://www. pmra-arla. gc. ca/english/pdf/fact/fs_pestreg-e. pdf accessed April 2005.

[40] Supreme Court of Canada Ruling:: *Hudson vs. Spraytech*, June 28 2001.

[41] Arya, N. (2005). Pesticides and Human Health: Why Public Health Officials Should Support a Ban on Non-essential Residential Use, *Canadian Journal of Public Health*, March-April, page 89-92.

In: Crop Spraying and the Health of Residents... ISBN: 1-60021-391-X
Editor: The Royal Commission, pp. 17-45

Chapter 2

PESTICIDE SPRAYING AND HEALTH

INTRODUCTION

2.1. The European Community (EC) Directive concerned with agricultural pesticides (Plant Protection Products Directive 91/414/EEC) specifies that 'when properly applied for the purpose intended ... agricultural pesticides should have no unacceptable influence on the environment in general and, in particular, no harmful effect on human or animal health or on groundwater'. [1]

2.2. The key issue raised throughout our study has been whether the health of residents and bystanders has been adversely affected by the spraying of agricultural pesticides.

It was clear from the consultation on buffer zones carried out by the Department for Environment, Food and Rural Affairs (Defra) in 2003, and from representations made directly to us, that a number of residents and bystanders think that this is the case. They are ill and they attribute their illness to pesticide spraying. We felt it appropriate to receive their evidence at first hand by visiting some of them in their homes.

2.3. Whether pesticides affect the health of residents and bystanders is an issue that has been raised as a concern for some time. In 1987 the then House of Commons Agriculture Select Committee reported on the effects of pesticides, stressing the need to respond to public concern about their possible effects on human health. [2] It concluded that assurances that the regulatory system is effective were required, that the system is seen to be effective, and that there is an unequivocal acceptance on the part of the agrichemical industry of the burden of proving that their products are safe beyond reasonable doubt. A further report on pesticides and health by the British Medical Association (BMA) in 1990[3] revealed the incompleteness of existing knowledge concerning the effects of pesticide exposure on human health, and the lack of a central government-led strategy governing the use of pesticides. Nearly two decades since the publication of these reports, some members of the British public continue to express dissatisfaction with the level of protection from pesticides and claim ill health as a consequence of their exposure. Those reporting ill health feel that there has been a lack of action by the government and/or medical profession on resident and bystander exposure. As a result many feel that they have on the whole been disregarded or even obstructed in gaining recognition of their concerns.

2.4. In this chapter we discuss the epidemiology and understanding of the health effects attributed to pesticides. We examine the toxicology data that underpin the approval of pesticides for agricultural use. Finally, we discuss the systems that are currently in place in the UK for residents and bystanders to report the ill health effects they attribute to pesticide spraying, both to medical professionals and to the regulators.

HEALTH EFFECTS ATTRIBUTED TO PESTICIDES

2.5. As discussed in chapter 1, it is difficult to estimate the numbers of people who feel that they suffer ill health related to pesticide exposure. Depending on the source, numbers vary from around a thousand to far fewer, and not all reported cases are related to agricultural exposure. On the other hand, not all cases may be reported or recorded. But on any basis the numbers appear to be a small proportion of the total number of people living close to sprayed fields, which are estimated at between 1 and 1. 5 million (1. 22).

2.6. To gain a better insight into the range of acute and chronic health effects that some members of the public are attributing to pesticide spraying, a Commission Member supported by the Secretariat visited the homes of 13 residents and bystanders and their families. We had no illusion that these visits would establish whether pesticides had actually damaged their health or not, but they were undertaken so that those affected and their families could relate their medical histories, which they believed were connected to pesticide exposure, and discuss the reactions of general practitioners and specialists, local farmers and neighbours.

2.7. The residents and bystanders visited lived in Lincolnshire, Norfolk, Essex, Devon, Hampshire, Gloucestershire and the Isle of Wight, some of the areas of the UK where intensive arable agriculture is practised. In addition, residents and bystanders were able to discuss their experiences of registering their concerns with local and central government bodies, in particular the Health and Safety Executive (HSE).

The Commission also received oral evidence from seven concerned individuals during its meeting in Cambridgeshire in February 2005. This was supplemented by over 50 submissions of written and video evidence concerning ill health and agricultural pesticide exposure, and in addition we had access to the submissions to the PSD consultation on Buffer Zones (1. 24).

2.8. The oral evidence that was collected from a small number of individuals during the visits contains valuable qualitative data, which allowed us to develop a better understanding of the wide range of health concerns reported by residents and bystanders. Evidence from the visits also provides pointers as to where further studies should be carried out.

Acute Effects Immediately Following Pesticide Spraying

2.9. The evidence from the residents and bystanders visited identified a series of well-defined acute symptoms immediately following pesticide spraying. These include upper and lower respiratory tract irritation, eye irritation, skin rashes, headaches and, in susceptible subjects, asthma attacks. We were also made aware of acute symptoms, that were less clearly defined, including confusion, loss of short-term memory, impaired cognition, dizziness and shortness of breath. Several of the residents and bystanders specified that these effects occurred in cycles associated with periods of pesticide spraying.

Chronic Health Effects

2.10. Residents and bystanders attributed a range of chronic health effects to crop spraying, some of which followed and some of which were unconnected with acute symptoms.

These included Parkinson's-like tremors and motility problems, sensory and motor neuropathies, allergic reactions (including skin rashes and angioedema), liver disorders and asthma. We encountered one case each of microphthalmia and polycystic ovary syndrome in the offspring of mothers who claimed they had been exposed to pesticide spraying, although in neither case had a causal link to pesticides been established. We also encountered well-defined disorders of the immune system that included anti-glomerular basement membrane disease (Goodpasture's syndrome), systemic lupus erythematosus (SLE) and mixed connective tissue disease, which residents and bystanders either attributed directly to pesticide exposure or suggested that pesticides had exacerbated these specific established diseases.

2.11. In evidence from residents and bystanders, general practitioners and non-governmental organisations, we encountered concern over the possibility that pesticide spraying in rural areas might be linked to clusters of cancers. These included lymphomas and leukaemia. We also encountered the view that pesticides might have potential endocrine disrupting properties that could lead to cancers, such as breast cancer.

We have not been able to ascertain whether the incidence of cancers in areas where pesticides are used is significantly above the expected rates for those populations, as there do not appear to have been systematic studies of these particular groups in the UK.

2.12. Some of those who had experienced exposure to pesticides suffered from multisystem and multisymptomatic disorders, which have been variously grouped under the terms of chronic fatigue syndrome (CFS) or myalgic encephalomyelitis (ME)[1] (appendix F) and multiple chemical sensitivity syndrome (MCS) (appendix G). The perception that there were clusters of chronic fatigue syndrome within small residential communities living near fields subject to frequent crop spraying was also reported to us, but again we have no evidence to link these reported clusters to pesticide spraying.

2.13. It became clear during our visits that those who considered their ill health to be caused by pesticide spraying had experienced and continued to experience genuine symptoms causing appreciable distress. Many of these individuals were very upset with the response from their general practitioners or consultant specialists. This led to multiple referrals to hospital consultants. A high proportion of residents and bystanders had resorted to tests and treatments at their own expense by complementary and alternative medical practitioners in the private sector. Residents and bystanders often also reported difficulties when trying to have their medical cases examined by any official government body (2. 77-2. 86).

2.14. Many of those who felt they had been affected by pesticides had become anxious and at times depressed, and their experiences often stretched over many years. There was evidence that irrespective of any direct effects that individual chemicals or their combinations might have on a person's medical condition, repeated spraying adjacent to their homes and outside their control increased their level of concern.

---

[1] For the purpose of this report, ME and CFS have been classified as the same condition, as was defined in the report of the CFS/ME working group:report to the Chief Medical Officer of an independent working group. January 2002.

2.15. Some of the medical complaints that residents and bystanders have told us about may have been associated with possible exposure to older pesticides, some of which have subsequently been withdrawn on account of toxicity or environmental persistence.

The fact that some pesticides initially regarded as being 'safe'have been subsequently withdrawn[2] has contributed to mistrust of the system and in the assurances given for the current generation of pesticides. However, we do recognise that over recent years there have been improvements in the selectivity of pesticides for their intended target with reduction in their broader toxicity and improvements in their environmental properties.

Mechanisms of Action of Pesticides and Possible Targets in Humans

2.16. There is clear evidence that chemicals used in pesticides do have biological effects in humans, which could explain disordered physiological processes that are considered to underlie some of the disease states observed in those reporting ill health. There are many examples of very similar molecular targets shared by target organisms and humans. One widely quoted example is the inhibitory effects of organophosphate pesticides, which have well-defined acute and chronic neurological effects in humans.

2.17. Some enzyme and other protein targets for pesticides, especially those for insecticides, have orthologues (evolutionarily related proteins) in humans. Organophosphate and carbamate pesticides that bind the enzyme acetylcholinesterase in insects[4] also bind to and inhibit the equivalent human enzymes. Acetylcholinesterase is an important enzyme that controls the level of acetylcholine, a neurotransmitter of nerve terminals, and is a molecule common to insects, humans and many other organisms. Although only 36% of the amino acid sequence of Drosophila (insect) acetylcholinesterase is in common with the human enzyme, their 3-dimensional structure are similar, and their enzymic active sites closely overlap. [5] Binding of an insecticide to the enzyme in humans is likely to lead to neurotoxic effects. Chronic insecticide exposure may have substantial deleterious effects including psychiatric symptoms and delayed neuropathy. [6]

2.18. However, the picture may not be as straightforward as that described in the previous paragraph. In an evaluation of the chronic neurotoxic effects of organophosphorus esters, [7] it has been found that, in addition to regulating acetylcholine levels in the peripheral and central nervous systems, acetylcholinesterases are present in virtually all mammalian cell types. They appear to have the function of preventing acetylcholine from acting as a hormone and localising its effect. [8] Non-neuronal acetylcholine is involved in many possible basic cell functions, including mitosis, cell differentiation and regulation of immune function. [9] These functions could provide further 'off-target'sites for insecticides that interfere with acetylcholinesterase.

2.19. As a further example, in a study to investigate chronic illness as a result of low-level exposure to synthetic pyrethroid insecticides, the London School of Hygiene and Tropical Medicine concluded that in the absence of decisive data on the effects of low-level exposure to pyrethroids, no conclusions could be drawn on whether or not it may lead to

[2] For example, under the EC Directive 91/414/EEC the agricultural organochlorine pesticide lindane has been withdrawn, on the basis that evaluation has identified concerns with regard to the safety of this active substance, in particular with regard to operators' exposure, its fate and behaviour in the environment and effects on nontarget organisms (2000/801/EC:Commission Decision of 20 December 2000).

chronic illness in humans. However, the study also emphasised that little work has been carried out to investigate chronic effects and further studies were necessary. The study recommended that properly designed neuro-behavioural studies on groups with long-term exposure to low doses of synthetic pyrethroids should be conducted. [10]

Epidemiology

2.20. Epidemiology is one of the ways in which scientists seek to investigate the distribution and determinants of disease frequency in human populations. [11] In the absence of toxicological data on humans or on some of the chronic and multisystem diseases experienced by residents and bystanders, there will always be an element of uncertainty about the risk that is posed by any pesticide. We agree with the view of Professor Coggon, Chairman of the Advisory Committee on Pesticides (ACP), that risk assessment needs to be backed up with sound epidemiological studies and health surveillance in order to investigate the evidence of any adverse effects in the public. [12, 13]

2.21. Since many of the diseases described by residents and bystanders, such as certain cancers (e. g. breast cancer), are common, it is important to determine whether there is a higher incidence of these diseases in residents and bystanders exposed to pesticide spraying than is found in the general population. Such analyses need to be carried out carefully with rigorous attention being paid to the end-points measured, the study power and the statistical analysis. These have created great difficulties in analysis of epidemiological studies in obtaining statistically significant results.

2.22. A major challenge associated with epidemiological studies of pesticides lies in the very considerable difficulty of quantifying exposure and identifying the particular pesticides or co-formulant mixture concerned. Not least because each pesticide acts in a different way. For pesticides, exposure is often evaluated retrospectively and through surrogate measures such as job description or the crop grown, [14] or by sampling individuals living at varying distances from fields known to have been sprayed with pesticides.

2.23. As a result, studies attempting to relate the use of pesticides in agriculture to a particular health effect are frequently limited to occupational groups such as farmers and commercial pesticide sprayers. This is eminently sensible as workers are exposed to higher levels of pesticides, but they are not representative of the general population as they are usually healthy adults and predominantly male. In contrast residents and bystanders include vulnerable groups, such as the elderly, infants and children, those with existing diseases and those at all stages of pregnancy. Residents and bystanders may be exposed to greater amounts of pesticide than has previously been recognised and also via different routes (chapter 3). Unlike residents and bystanders, workers applying pesticides to farmland are advised by the Green Code[15] (5. 70-5. 71) to wear basic personal protective equipment (PPE) when handling pesticides, as well as any further PPE that is advocated by the statutory label on the product. Workers using a boom sprayer to apply pesticides may also be sitting in a pressurised cab that isolates the driver from the outside.

2.24. Examples from other fields demonstrate that levels that might be safe for workers may be inadequate to protect the health of the wider population. For example, the Expert Panel on Air Quality Standards (EPAQS) has recommended a lower standard for general population exposure to nitrogen dioxide than the occupational exposure standard that has been set. This was based on epidemiological studies suggesting that adverse health effects may occur at exposures to lower concentrations of nitrogen dioxide than the occupational standard over longer periods of time. [16] Nevertheless the mechanism for setting occupational exposure standards is not as rigorous as for pesticides. Exposure assessment is further discussed in chapter 3.

2.25. In agricultural situations many chemically different pesticides will be used over time and during the growing season making it difficult to ascertain which pesticide, if any, might be implicated in a particular health effect. A further complication is that some chronic diseases (in particular cancers) have long latent periods of years or decades before clinical manifestation. This makes any association between pesticide exposure and the presentation of a disease difficult to establish.[3]

2.26. Chemicals that produce a 'unique'reproducible disease in humans are the most likely to be identified, particularly if they produce a similar effect in animals. Conversely, if a pesticide increases the incidence of a very common human disease, it might be impossible to identify it as the cause. [17] Despite these difficulties, it should be feasible to devise epidemiological studies to shed further light on the possible effects of pesticides on the health of residents and bystanders. Organisations such as the Small Area Health Statistics Unit (SAHSU) at Imperial College of Science, Technology and Medicine, London, are well placed to undertake investigations into alleged clusters of disease in rural areas. Studies of this kind were discussed in more detail in the Commission'sTwenty-fourth Report, Chemicals in Products.

Reviews of Epidemiological Studies

2.27. Recently several reports have attempted to analyse the epidemiological literature with regard to particular health effects of pesticides.

2.28. In 2004, the Institute for Environment and Health published a review of Parkinson's disease and the potential role of pesticides, commissioned by the Pesticides Safety Directorate (PSD) on behalf of Defra. [18] Taking into account epidemiological and toxicological evidence on specific compounds and mechanisms, the report concluded that there was sufficient evidence to suggest that exposure to pesticides might be implicated in an increased incidence of Parkinson's disease. However, the existing body of evidence was considered insufficient to establish causation for any specific pesticide or pesticide combination. [19] A recent large study involving five European countries, the 'Geoparkinson study', aimed to investigate whether exposure to chemicals increases the risk of developing Parkinson's disease and to determine the role of a number of genes in modifying the risk. The study has established an association between pesticide exposure and Parkinson's disease that exhibited an exposure—response relationship. [20, 21]

2.29. The Department of Health's Committee on Carcinogenicity of Chemicals in Food, Consumer Products and the Environment (COC) was asked to review the available research,

[3] Examples of the latency of cancers caused by environmental exposure are arsenic poisoning with a latency of up to 20 years (World Health Organization website http://www. who. int/water_sanitation_health/dwq/wsh0306/en/index2. html accessed March 2005) or in the case of exposure to asbestos of the order of 10-50 years (World Health Organization website (http://w3. whosea. org/en/Section23/Section1108/Section1835/Section1864_8658. htm accessed March 2005).

including epidemiological studies, to identify if there were any chemical exposures that might be associated with prostate cancer, and to consider the evidence for an association between farmers, farm workers and pesticide spray operators and increased risk. The Committee concluded that there was some evidence of an association between farmers/farm workers, exposure to pesticides and increased risk of prostate cancer, and concluded on the basis of this that the future published literature should be carefully monitored for further studies. [22] In 1995, 1999 and again in 2003, COC reviewed the epidemiological literature on specific organochlorines and breast cancer, and concluded that the available evidence did not indicate a clear association or significant risk. [23]

2.30. The Ontario Group of Family Physicians, a small team mainly composed of family physicians and graduate students but with epidemiological expertise, reviewed the literature on the human health effects of pesticides published in the period 1992-2003.

The pesticide review was funded by the Laidlaw Foundation as the first step in upgrading the Ontario Group of Family Physicians'brochures and educational programs on the impact of pesticides on public health. The criteria for choosing the papers were that the studies had to be systematic in their approach, and be written in English, French, Spanish or Portuguese. The searches were done using the PreMedline, Medline, CancerLit and LILACS (Spanish-language) databases. This review excluded literature on organochlorines, as most of these chemicals have now been classified as persistent organic pollutants (POPs)[24] and have been phased out of use as pesticides. Thirty review papers (synthesis of primary sources) were selected and categorised according to health effects: cancer, genotoxicity, immunological, neurotoxicity, reproductive and other. Of primary research studies, a total of 109 papers where the target disease was cancer and 156 non-cancer papers met the criteria for consideration. [25]

2.31. The report found that many of the studies showed statistically significant positive associations between pesticide exposure and solid tumours including those involving the brain, prostate, kidney and pancreas. Relationships were also found between 2, 4-D (2, 4-dichlorophenoxyacetic acid) and related pesticides and non-Hodgkin's lymphoma (NHL), pesticides and leukaemia, as well as consistent effects linking pesticide exposure to disorders of the nervous systems. Reviewed studies suggested that there were some statistically significant associations between occupational exposure to agricultural chemicals and adverse reproductive effects, including birth defects, foetal death and intra-uterine growth retardation. [26] The review attracted media attention, in part because of current interest across Canada in by-laws to prohibit or reduce the cosmetic use of pesticides. [27]

2.32. In a critical response to the Ontario review, the UK Advisory Committee on Pesticides (ACP) concluded that 'the report was seriously flawed, did not raise any concerns about pesticide safety that were not already being addressed, and did not indicate any need for additional regulatory action in the UK'. According to the ACP, [28] flaws in the study included failure to take account of all of the relevant epidemiological evidence, and biases inherent in the way in which papers were selected for inclusion. The ACP felt there was inadequate attention to exposure characteristics and relevant toxicology when interpreting reported associations and superficial synthesis of the evidence, which in the ACP's view inadequately explored the impact of the strengths and weaknesses of individual studies. The ACP later published a minority statement from one member on the Ontario review, stating that 'there is a range of views on the conclusions of the Report, and I do not agree with the statement issued by the ACP. Alternative views are relevant and in my view have not yet been given adequate consideration. '[29] We discuss the full range of opinions in scientific advisory committees (5. 38-5. 43).

2.33. We sought an independent review of the Ontario study by Dr Michael Burr, Reader in Epidemiology at the University of Wales College of Medicine. [30] He made the overall comment that the subject is of considerable potential importance and of interest to the general public and health professionals. The rationale for conducting the Ontario study was clearly explained, and the document was well written and easy to follow.

A number of limitations were however pointed out. There was no meta-analysis (systematic quantitative summary) of the primary studies, and the treatment of review papers appeared unclear and inconsistent. The authors had insufficiently addressed the issue of publication bias, and the review seemed to over-interpret the findings, given the limitations of the relevant studies. Strong conclusions were being drawn from evidence that was of rather weak quality. Dr Burr concluded that it was difficult to assess the likelihood and strength of causal effects in the various associations reported.

2.34. In summary, there has been a large amount of scientific literature produced on the question of pesticides and chronic health. Some studies indicate a link, but many others have failed to establish any significant link. A number of reviews criticise the methodology or findings of other studies. The field is vast, complex and controversial.

A robust and comprehensive critical review of the health-based literature or the use health-based evidence in an attempt to attribute causality would take a large amount of resources. In concluding his review of the Ontario report, Dr Burr made a strong case for setting up a large, definitive study that incorporates good measures of exposure and we agree that such a rigorous study is required.

Multisystem Disease

2.35. As well as acute and chronic ill health effects, particularly cancers, residents and bystanders have linked symptoms associated with multisystem, multisymptom diseases to pesticide exposure, and this is an area where there is a particular paucity of information. While we acknowledge that debate continues into the pathophysiological mechanisms behind these disorders, we recognise that they are likely to be of various origins and that several different mechanisms are likely to interact in a complex way leading to the development of symptoms. International scientific and medical bodies are increasingly acknowledging these disorders, although there remains a strong body of opinion that most of the symptoms can be explained by the physical manifestation of underlying anxiety and related psychiatric disorders. [31, 32] We consider some of these multisystem diseases here, in particular chronic fatigue syndrome (CFS) and multiple chemical sensitivity (MCS), and discuss some new evidence and techniques for investigating such illnesses. Brief descriptions of CFS and MCS are provided in appendices F and G.

2.36. After many years of debate and divided opinion over the existence of CFS, in 2002 an independent working group in the UK under the chairmanship of Professor Allen Hutchinson produced a report for the Chief Medical Officer which formally recognised the existence of CFS as a chronic and debilitating illness deserving specific forms of treatment. [33] Based on the evidence available to them at the time the group concluded that the balance of evidence indicated that environmental toxins were not a common or widespread trigger. A report in 2002 by the Danish Ministry for the Environment reviewed the epidemiological literature and concluded that MCS does exist, even though the exact mechanism associated with the disease is not known and there are no demonstrable organic or functional changes. [34]

2.37. Although frequently labelled as a psychiatric disorder, recent epidemiological and pathophysiological evidence provides an increasing case for the existence of MCS as a distinct clinical entity (appendix G). Beyond these syndromes of uncertain origin we encountered other disorders such as chronically impaired cognition and memory, sleep disturbance and lack of concentration that were common among those attributing their symptoms to pesticide spraying. We also heard evidence from general practitioners that all of these disorders are presented not uncommonly in primary care, and that pesticide spraying may be only one of many different factors contributing to their pathophysiology. [35] It has been pointed out to us that multisymptom disorders are not simply a function of the twenty-first century. For example, 'neurasthenia'was widely reported in the literature of the late nineteenth century, [36] although in this case, despite many theories and claims, no organic basis for the disorder was ever established. The medical profession eventually rejected the term in the UK, although the term remained in other parts of the world, and still exists in the World Health Organization's International Classification of Diseases. [37]

2.38. In the case of CFS there is some evidence for the involvement of susceptibility genes, and immunological and neurological pathways. [38] Very few patients with multisystem disorders have been studied using modern investigation techniques such as magnetic resonance spectroscopy (MRS) of the brain (appendix H), although such procedures have proven successful in altering attitudes in the US to the existence of Gulf War Syndrome, [39] which shares with the subject of our enquiry similar multisystem disorders (appendix I). The evidence from the US 2004 report by the Research Advisory Committee on Gulf War Veteran's Illnesses supports further research and the collection of information from people with multisystem disorders, who feel that the cause may potentially be exposure to pesticides.

2.39. For over 13 years Gulf War Syndrome was not recognised as a clinical entity. It is only in the past six months within the US that evidence has been reported to suggest otherwise. We feel that this example strengthens the case for a full investigation into possible links between pesticide spraying and ill health in residents and bystanders.

It is our view that a new research programme involving the Research Councils and the Department of Health should be undertaken to address the lack of knowledge on the possible chronic health effects of pesticides.

TOXICOLOGY

Introduction

2.40. The current regulatory approach to the management of potential risks from pesticides involves a process of risk assessment. This is developed from risk characterisation, which integrates the information assembled from hazard assessment (toxicity testing) with the route into the body (exposure assessment), using a combination of qualitative and quantitative information, together with information regarding uncertainties, to produce conclusions on the likelihood of adverse effects occurring in the exposed population or individuals. [40] Hazard assessment is described in further detail in our Twenty-fourth Report, *Chemicals in Products*.

2.41. In this section we look at the principles of the hazard (toxicity) assessment that are used to calculate acceptable exposure for the public, while the rest of the risk assessment process (the exposure assessment) is described in chapter 3. As pointed out in our Twenty-first Report, *Setting Environmental Standards,* it is not possible to define an acceptable level of risk using purely scientific methods as communities will vary in what they choose to regard as acceptable risks, and it has been argued that a risk can only be described as acceptable if the public regards it as such. [41] Acceptability of a risk will also depend on the confidence that the public has in the process that informs risk assessment. [42] We examine this acceptability of risk further in 5. 33-5. 37.

2.42. Hazard assessment is based largely on data collected from animal toxicological testing.

The testing required is set out in detail in the EC Plant Protection Product Directive (91/414/EEC). In the UK, the Pesticides Safety Directorate (PSD) assesses the toxicological evaluation undertaken by a manufacturer seeking approval for a new agricultural pesticide. Pesticide testing differs from pharmaceutical testing. For pesticides no toxic effect is permitted, while for pharmaceuticals, a balance between efficacy and safety is viewed as a 'trade-off'. Pharmaceuticals are tested in human clinical trials to identify both positive outcomes of treatment and harmful side effects before they are generally prescribable. There will also be continued careful monitoring of health impacts in humans receiving the specific drug (pharmacovigilance) once they have been authorised for use. Apart from *in vitro* tests pesticides are usually only tested *in vivo* in animals.

Testing Paradigms

2.43. Toxicology is a branch of science studying the harmful effects of chemicals on living organisms (their intrinsic hazard). Toxicological testing is predominantly undertaken on pure-bred animals, such as the Wistar rat, for which there are good background data. [43] Rodents have proved an excellent model of mammalian physiology, but are simply a tool used by risk assessors to predict what might happen in humans. Uncertainty factors are incorporated into the process (2. 49-2. 50), but these may not be relevant to the complex problems encountered by those who complain of pesticide related problems. The human population has relatively wide genetic variations, different age groups, gender and states of health.

2.44. For pesticides, the toxicological evaluations required are specified in the EC Plant Protection Products Directive (91/414/EEC). These tests are all conducted according to international guidelines (e. g. from the Organization for Economic Co-operation and Development (OECD) and principles of Good Laboratory Practice (GLP)).

The Department of Health's Health Products Regulatory Agency's Good Laboratory Practice Monitoring Authority (GLPMA) regularly inspects UK laboratories. The manufacturer must conduct an extensive series of tests for all pesticides irrespective of their intended use, and submit a full database to the Regulator – the PSD in the UK –as a requirement of pesticide registration. [44] These toxicological evaluations are undertaken principally for a single active ingredient, with more limited tests for the acute effects from the formulation or finished product. [45] Regulators then decide whether there is potential for interaction between active ingredients as part of the risk assessment process. [46]

2.45. A series of well-defined tests is used to elucidate any acute effects from exposure to both active ingredient and pesticide formulation. In addition, systemic, neurological, reproductive, developmental and oncogenic effects resulting from either a single or lifetime exposure are investigated, but these tests are carried out on the active ingredient only. Animals are exposed to high levels of a single pesticide active ingredient from hours to years, and the gradation of effects resulting from increasing doses is used to establish a dose–response relationship and no observed adverse effect level (NOAEL) in the animals. [47]The minimum testing requirements are outlined in table 2. 1.

Table 2. 1. Minimum animal testing requirements of EC Directive 91/414/EEC for plant protection products[48]

TEST	ACUTE	CHRONIC AND LOCAL	APPEARANCE / OBSERVED
ACUTE AND LOCAL	Oral, dermal and inhalation routes for evident toxicity.	Skin and eye irritation. Skin sensitisation.	Evident toxicity, and local irritation and sensitisation.
SYSTEMIC	Acute for systemic toxicity (sub lethal), in the rat and the dog. Sub-acute e. g. 28 days.	Sub-chronic – 90 days Chronic – 1 year	Shape and structure of all major organs and organ systems, growth, integrated function and ability to respond to stimulation.
NEUROLOGICAL	Acute	Sub-chronic – 90 days in the rat.	Shape and structure and in-depth assessment of nervous system, growth, integrated function and ability to respond.
	Sub-acute e. g. 28 days in the rat and the dog.	Chronic – 1 year in the rat.	
DEVELOPMENTAL		Dedicated developmental neurotoxicity study in the rat (not routine); post natal development. *In utero* development in the rat and the rabbit.	Shape and structure (external and internal of the foetus including the skeleton). *In utero* survival; growth and development of the whole organism.
REPRODUCTIVE		Multi-generation reproduction study in the rat.	Mating behaviour; Fertility and fecundity; Life-stage susceptibility; Generational effects; Survival and growth; General and sexual development; Sperm analysis.
ONCOGENICITY	Short-term tests for genotoxicity, including in vitro mutagenicity tests (bacterial assay for gene mutation, test for clastogenicity in mammalian cells and test for gene mutation in mammalian cells).	Lifetime bioassays in the rodent – rat and mouse.	Survival; Tumours; General histopathology.

2.46. The many observations that are collected during toxicological testing are analysed for unexpected patterns of pathology. The end-points observed include any physical changes in the animal (pathological histology), and unusual variations in growth and observed behaviour. Changes in behaviour would include reflex response to stimuli used to identify potential neurological effects in animals as a surrogate for effects such as altered consciousness, fatigue and headaches in humans (symptoms not infrequently described by residents and bystanders). While such observations of behaviour cannot be quantified through specific tests (as it would be possible in the case of tumours), [49] a limited amount of information on quantification can be obtained by scaling the symptoms and by reporting the number of animals with symptoms at each dose.

2.47. On our visit to Syngenta, we were impressed with the testing regime undertaken for pesticides and in particular the development of new tests that are not currently a requirement of the regulatory bodies. While neurological testing is routinely undertaken (table 2. 1), we were also provided with evidence of a newly developed functional observation battery (FOB) of tests, where each rodent is removed from its cage a number of times during the testing period and examined for changes in general health and physical assessments of landing foot splay, muscle weakness and sensory perception (tail-flick test). Following the FOB, locomotor activity is monitored using an automated activity recording apparatus. The instrument records small and large movements as an activity count. Both tests are undertaken by personnel who are experienced in handling these animals and are able to ascertain if there are any unusual changes in neuro-performance that might be caused by exposure to pesticides. [50]

2.48. Currently, there are no animal models available to reflect all of the chronic ill health effects that many residents and bystanders have associated with pesticide spraying, specifically multisystem, multisymptom disorders such as CFS (appendix G), which may result from long-term exposures to low doses of pesticides, but new models are being developed that hold promise. We are aware of some new animal models that reproduce some aspects of CFS, such as cholinergic hypersensitivity and fatigue, [51, 52] that deserve further study. It is important to ensure that symptoms and syndromes that do not fit within the traditional toxicological end-points observed are not ignored simply through lack of knowledge. This applies to the study of multisymptom illness regardless of origin. In effect, there is a lack of understanding and, therefore, good input data on no-effect levels for many illnesses that residents and bystanders complain of. Regulatory exposure assessment is further addressed in chapter 3.

Uncertainty Factors

2.49. Laboratory animals are studied as a whole system model to gain an understanding of how humans might respond to the same chemical (over a range of outcomes as in table 2. 1). The no observed adverse effect level (NOAEL) is derived from the maximum exposure to a chemical residue that regulators are confident will not result in an adverse effect. At concentrations above the NOAEL there may be an effect or a specific toxicity end-point may be observed. Genotoxic carcinogens (for which it is assumed there is no threshold) are not approved for use as pesticides.

2.50. In the risk assessment process, an acceptable exposure level in humans, the acceptable operator exposure level (AOEL), is calculated from the NOAEL. The AOEL is calculated by dividing the NOAEL in animals by an uncertainty factor of 100 (figure 2-I).

This uncertainty factor incorporates a factor of ten for inter-species variability (animals to humans) and a further factor of ten for population variability (to protect vulnerable sub-groups of the population such as children and the elderly). [53] In the absence of a no effect or NOAEL value, the lowest observed effect level (LOEL) from toxicological tests is used together with an increased uncertainty factor, to calculate the AOEL.

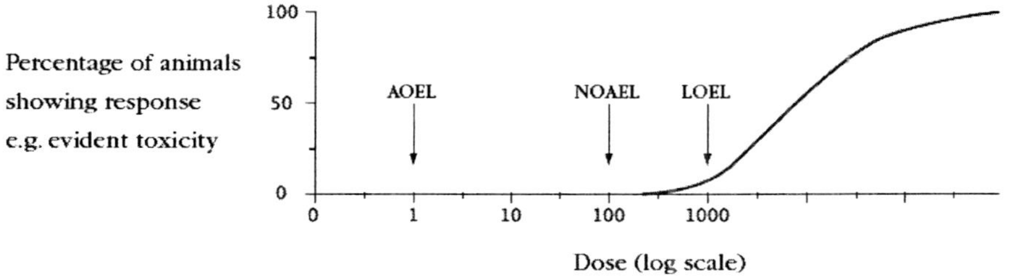

Figure 2-I. An example of a dose–response curve, showing the relative positions of the lowest observed effect level (LOEL), no observed adverse effect level (NOAEL) and acceptable operator exposure level (AOEL)[54].

New Technologies and Techniques

2.51. We are concerned that toxicological testing as currently undertaken focuses on specific endpoints and does not cover many of the chronic multisystem effects described to us by members of the public. We recognise that it will be difficult, but not impossible, to devise testing arrangements that address these concerns, but believe that the element of uncertainty inherent in using animal models might be reduced by the development of new integrated and molecular based technologies, [55] such as the use of toxicogenomic methods or human cell culture models and the development of animal models of multi-symptom/multi-system disease.

2.52. Toxicogenomics is the application of knowledge of the genes associated with disease states to the study of the toxicology of chemical and physical agents. [56] It combines the studies of genetics, genomic-scale mRNA expression (transcriptomics), cell and tissue-wide protein expression (proteomics), metabolic profiling (metabolomics) and bioinformatics, along with conventional toxicology information, in efforts to understand the modes of action of chemicals and the potential role of gene–environment interactions. [57]Toxicogenomics is a tool in development, but it has the potential to inform and improve risk assessment in the future, [58] especially if used in conjunction with altered gene expression identified from affected individuals. Toxicogenomics was discussed in further detail in the Commission'sTwenty-fourth Report, *Chemicals in Products*. The Committee on Toxicity of Chemicals in Food Consumer Products and the Environment (COT) has recently conducted a full review of the use of toxicogenomic methods in toxicology and has noted the potential use of these methods in the future. [59]

2.53. Under stringently controlled experimental conditions, human cell culture models can be used to re-create human cellular function in an in vitro environment and to demonstrate mechanistic, physiological response to a pesticide, as has been done with benomyl. [60] This could reduce the need for animal models and animals for experimentation and deserves further exploration by those involved in pesticide regulation. This can reduce the need for animal models and animals for experimentation. However, there is a need to consider this in synergy with our previous comments (2. 35-2. 39), which indicate that better characterisation of these diseases is needed before models can be created to test for them.

MONITORING

2.54. The proper characterisation of exposure is a vital element in understanding the relationship between environmental contamination and adverse health effects.

Inadequate knowledge of exposure is a key weakness of epidemiological studies (2. 20-2. 26). [61]

2.55. The duration, magnitude and timing of exposure strongly influence the severity (or probability in the case of cancer) of effects. Acute exposure is defined as an exposure that has a duration of less than 24 hours, sub-acute exposure 24 hours to 3 months, while chronic exposure extends is defined as extending beyond 3 months.

Exposure in itself does not necessarily produce harmful effects, as these depend on the dose and intrinsic toxicity of the compound, as well as the route into the body and the sensitivity of the persons exposed. Relating reliable exposure information to biomarkers of pesticides (e. g. dialkylphosphates are a biomarker of organophosphate pesticides) in the body or to health effects that have been carefully monitored could help inform assessment of the risk to humans. [62]

Biomarkers of Pesticide Exposure

2.56. Data on levels of pesticide biomarkers in the general population could provide valuable information for assessing how any changes in pesticide usage are reflected in the exposure of the population, [63] noting that the presence of a pesticide in a sample does not necessarily mean that there will be an adverse effect in that person. Baseline data on levels of exposure in the population would allow comparison with biomarker levels in an individual subject and provide an understanding of whether the level is unusual and in a range that might lead to an adverse effect. This information could be compared to symptoms of ill health and analysed for trends. We were surprised to find that no efforts have been made to establish a database of baseline information for agricultural pesticides that are commonly used in the UK. The principle behind comparing an individual's level with the population norm (± variance) is an entirely standard method of proceeding in many areas of clinical diagnosis.

2.57. Baseline information is being collated in other countries, notably in North America and Germany, [64, 65] and could be used as a framework for information that could be collected in the UK. There have been a few smaller-scale studies of note in the UK, discussed below, showing that the collection of baseline biomarker information can be done relatively easily.

US Studies

2. 58 In the USA, the National Health and Nutrition Examination Survey (NHANES) is a periodic survey conducted by the National Center for Health Statistics (NCHS) of the Centers for Disease Control and Prevention (CDC). Data obtained from national probability sample surveys provide important information on the prevalence of various health conditions and distributions of physical and biochemical characteristics in the general population, [66] and can be used by physicians to determine whether a person or group of people has an unusually high exposure to the substances measured.

Chemicals or their metabolites were measured in blood and urine samples from selected participants in the NHANES survey. During 1999-2000 116 environmental chemicals were measured, including dialkylphosphate metabolites of organophosphate pesticides, organophosphates, organochlorine pesticides, carbamate pesticides, herbicides and pest repellents. [67] CDC's *Third National Report on Human Exposure to Environmental Chemicals*, published in July 2005, includes an expanded range of general chemicals and also pyrethroid pesticides. This work could be used as a model for how a national baseline study might be set up in the UK, for both environmental chemicals and pesticides.

2.59. In the USA, a recent state court decision has mandated the state of Washington to implement a plasma cholinesterase (ChE) surveillance program for all agricultural workers, as a biomarker for specific insecticides. Currently, only Washington and California conduct mandatory testing of farm workers who mix and spray N-methylcarbamate or organophosphate pesticides (which both act as acetylcholinesterase inhibitors)[68] for 30 or more hours in any consecutive 30-day period. [69] Monitoring of pesticide handlers on a monthly basis can detect over-exposure and prevent illness. [70] The initial results indicate that an unexpectedly high 21% of workers had sufficiently low levels of ChE to put them into the 'at risk' category. [71] This work demonstrates the power of using biomarkers to monitor people who may be exposed to pesticides through their work.

UK Projects

2.60. We received evidence from Professor Kevin Jones of Lancaster University about the work he undertook for a WWF biomonitoring survey. [72] One hundred and fifty-five blood samples were taken in 13 locations in the UK, and analysed for 78 chemicals including 12 organochlorine pesticides. The pesticides examined included the older types no longer in use in the UK and so aimed to report on long-term persistence, and it was noted that modern pesticides do tend to be less persistent and bioaccumulative. [73] HCB (hexachlorobenzene), p, p'-DDE (dichloro-diphenylichloroethylene, a major metabolite of DDT), p, p'-DDT (dichlorodiphenyltrichloroethane) and _-HCH (beta-hexachloro-cyclo-hexane) were the most commonly detected organochlorine pesticides and were dominant in almost all samples. [74] The relevance of this study is that it demonstrates that baseline information can be collected nationally and relatively easily, especially for the more persistent and bioaccumulative pesticides.

2.61. As far as we are aware, no studies in the UK have attempted to measure biomarkers for pesticide exposure in residents and bystanders that might inform the risk assessment process. Some considerable time after the 1987 House of Commons Agriculture Select Committee report and the 1990 BMA report (2. 3) we find this surprising, given that the public's health had been identified in both reports as being potentially at risk.

A pilot study of biological monitoring of 30 rural bystanders has recently been commissioned as part of the research and development work of the PSD, although we understand that the study has yet to start. [75] This new work aims to compare the levels of a specific pesticide metabolite in the urine of adults and children with documented exposure to the chosen pesticide (the insecticide cypermethrin) and with the level of metabolite predicted from data obtained from the regulatory process. [76] This should provide a basis for assessing the reliability of the present exposure assessments. [77] We feel that such work should be started as soon as possible and then expanded to include a wider range of pesticides.

2.62. Further UK projects that could provide opportunities for biological samples collected for other purposes to be used to measures biomarkers of pesticide exposure[78] include the UK Biobank that will start in 2006 and aims to collect blood and urine samples from 500, 000 volunteers aged 45-69. [79] The Biobank study focuses on DNA-based genomics, but is an example of collection of biological samples on a large scale. Biological samples have also been collected in the Surveillance of Residues in Human Milk (SUREmilk) pilot study of 400 women in Yorkshire undertaken in 2002, which monitored contaminants in human breast milk. [80] In Germany, the Environmental Specimen Bank (ESB) provides another example of the collection of samples from human volunteers. Samples have been collected and archived since 1997, and are being used as a general trend indicator and for future analysis. [81]

2.63. An additional study which could be used to obtain baseline information is the Avon Longitudinal Study of Parents and Children (ALSPAC), an ongoing study of 14, 000 children from the Avon area around Bristol. The children who form the basis for this study were all born in 1991 and 1992, and have been followed from pre-natal existence onwards. The study aims to improve understanding of how physical, genetic and social environments interact over time, affecting health, behaviour and development. [82] It uses a variety of data collection methods including questionnaires, physical examination, health records and the collection of biological samples (which could be analysed for pesticide biomarkers). This information is being used to test specific hypotheses about the causes and prevention of childhood ailments and disorders. [83] The clear advantage of a longitudinal study is that the information collected could improve understanding of the impacts that pesticides may have on health over the long term. The ALSPAC study has already been used to collect information on the use of pesticides in homes and gardens, in order to understand the contribution that parental pesticide use might make to total pesticide exposure. [84] Another example of the potential of this cohort is a recent asthma study which suggested that frequent use of chemical-based products in the pre-natal period is associated with persistent wheezing in young children. [85]

2.64. We believe that it is important to develop national databases of information from sample surveys and large-scale longitudinal studies, so that baseline information on the exposure of the UK population to pesticides and other environmental chemicals can be derived and used for comparison with selected biomarkers in individuals who are concerned about their personal level of exposure. Such information could be used in epidemiological studies and surveillance to gather information about trends of symptoms and illnesses, which might be associated with a particular exposure level (2. 87-2. 96).

Recommendations

2.65. Based on the conclusions from our visits and our understanding of the biological mechanisms with which pesticides interact, it is plausible that there could be a link between bystander pesticide exposure and chronic ill health. We find that we are not able to rule out this possibility. We recommend that a more precautionary approach is taken with passive exposure to pesticides. The existing uncertainties indicate an urgent need for research to investigate the size and nature of the problem and any underlying mechanisms that link pesticide spraying to ill health.

2.66. Recognising the enormous, complex and diverse literature on ill health caused by pesticide exposure: we recommend a comprehensive systematic review of the literature on pesticide spraying and human health that takes account of the shortcomings of the Ontario Report.

2.67. Based on the recent findings in the investigation of Gulf War Syndrome (appendix I), which shares some common features with the subject of our report, we are concerned that similar modern tools of molecular and integrated science have not yet been used to investigate those who express adverse chronic health effects from pesticide spraying.

We recommend that an imaginative systematic approach is taken to apply both well validated as well as novel clinical investigative methods to those with chronic symptoms linked to pesticide spraying such as magnetic resonance spectroscopy (MRS) and gene and protein profiling.

2.68. Little is known about the body levels of pesticides in those living in localities where pesticide spraying occurs compared to those not so exposed. We recommend that the Health Protection Agency (HPA) (2. 86) and related organisations within the devolved administrations in Scotland and Wales collect population data on pesticides, their metabolites and biomarkers of effects that would provide a sound basis for exposure assessment and also could be used to establish a national database for monitoring.

2.69. Recognising that many of the medical problems that residents and bystanders link to pesticide spraying are complex and may not be easily modelled in the toxicology screening process of pesticides: we recommend that the private sector and universities be encouraged to develop new animal models that better reflect the chronic disorders experienced by residents and bystanders exposed to pesticide spraying.

Health Effects, Monitoring and Reporting

2.70. This section discusses how the suspected adverse health effects of exposure to pesticides in residents and bystanders are managed; the important role that primary care holds in dealing with patients who feel pesticides have caused their ill health, and the current arrangements for reporting and monitoring adverse health effects resulting from pesticides exposure in the UK.

Primary Care

2.71. General practitioners and hospital doctors are the first line of defence in the treatment of public health from the hazards of toxic chemicals in the UK. Primary health care professionals are usually the first contact for affected members of the public and are responsible for making referrals to medical specialists in the secondary care sector.

However, training of medical undergraduates in the basic principles of toxicology is given a low priority in medical schools. [86] In its 1990 report on pesticides, the BMA pointed out that clinical toxicologists were uncommon in the National Health Service (NHS). Since the report was published the situation has not improved. The BMA suggested in their report, that while not all medical practitioners need to be expert toxicologists, they do need to be aware of the possibility of disease resulting from exposure to toxic substances such as pesticides and other environmental chemicals, and to have more training and practical experience in the diagnosis and treatment of chemically-induced disease. We recognise the pressures on general practitioners and the medical training curriculum training, but feel that there is a need for sufficient training to enable general practitioners to at least recognise the symptoms that may relate to exposure to pesticides and other environmental chemicals, so that they are able to refer patients to a more specialised tertiary level for diagnosis and treatment.

2.72. Currently, all health care professionals have access to the National Poisons Information Service (NPIS) as a resource for toxicological information and expertise. The first point of contact with the NPIS is the use of an online computerised database on poisoning (TOXBASE). If further expert advice is required, a 24-hour emergency information line staffed by physicians and clinical scientists is available. This system is reliant on primary care professionals identifying the need to contact the NPIS system and thus tend to be of an acute nature, and requires details of the actual pesticide that has caused the ill health effect.

2.73. Currently, a significant proportion of residents and bystanders with health concerns that they attribute to pesticides refer to complementary and alternative practitioners to receive any advice and treatment (2. 13). There is a need for a clear referral pathway from primary care to NHS physicians adequately trained in environmental medicine, so that residents and bystanders do not have to feel that they are forced down the route of complementary practice in order to get treatment.

2.74. In the USA the *National Strategies for Health Care Providers:Pesticides Initiative* is attempting a strategic approach incorporating environmental health information into the education and practice of primary health care providers using pesticides as a model.

This long-term initiative is a project of the National Environmental Education and Training Foundation (NEETF) in partnership with the US Environmental Protection Agency in collaboration with a broad range of stakeholders including government agencies, health care provider organisations, and academic institutions. The aim is that primary care providers can, through education and training, identify exposures and work with experts to manage their patients'suspected pesticide-related illnesses and concerns. [87] Such a system could be used as a model for primary care, and for the diagnosis and treatment of environmental disease in the UK.

2.75. Pesticides have been a topic of concern in the UK for at least two decades (2. 3).

This is reflected by the Department of Health having issued the handbook *Pesticide Poisoning Notes for the Guidance of Medical Practitioners*, in both 1983 and 1996. The 1996 version was distributed to all general practices, hospital accident and emergency departments, Consultants in Communicable Disease Control (CCDC), medical schools and professional bodies, together with a press release and communication from the Chief Medical Officer. This second edition of the handbook was written to update information originally provided in response to the 1990 BMA report *Pesticides, Chemicals and Health*. [88]

2.76. The handbook was created to help physicians diagnose and treat acute pesticide poisoning, with advice on first aid and patient management. Chronic symptoms are covered in Appendix 6 of the handbook, where the text states that 'not infrequently patients complain that their chronic ill health is the result of exposure to environmental toxins, commonly pesticides, and hold this view with absolute conviction. The following protocol is one approach to dealing with them but clinical examination and extensive investigation usually fails to identify any pathological lesion. 'The protocol briefly describes the patient history that should be taken and some investigative tests that can be done, as well as advising the physician that they may wish to report the case to the relevant authority. [89] General practitioners receive large volumes of material, and a decade after the handbook was distributed it is likely that awareness of it is low. [90]

Reporting Public Ill Health Effects of Exposure

2.77. Currently, no single regulatory department or agency is responsible for dealing with public health effects resulting from exposure to pesticides that are used in agriculture (5. 4). One aspect of this is dealing with complaints from the public. The Health and Safety Executive (HSE) has the responsibility for protecting residents and bystanders from pesticide exposure arising from commercial farming, through the Health and Safety at Work Act 1974 (Section 3) and Control of Substances Hazardous to Health (COSHH) Regulations 2002 (as amended). [91]

2.78. The HSE has published a booklet for the public, which is also available on its website, informing people what to do if they feel that their health has been affected by pesticides. [92] The booklet states that the first thing to do is to seek medical advice immediately from a general practitioner or by going to hospital. There is thus an immediate focus on primary care practitioners as being able to diagnose and treat pesticide-related ill health. Subsequently, the booklet states that a report should be made in writing or by telephone; in the case of pesticide drift from commercial farming this would be to one of the HSE's Regional Offices.

2.79. Once a report has been filed, an HSE Inspector will investigate pesticide complaints involving alleged ill health effects. Depending on the outcome of this investigation, an inspector may give oral or written advice to, issue a notice to or, if appropriate and if there is enough evidence, consider prosecution of the farmer or pesticide sprayer. [93]

2.80. On completion of the investigation, the HSE Inspector will send details of reported cases to the Pesticide Incidents Appraisal Panel (PIAP). PIAP was established by the HSE to examine the information compiled from its investigations into incidents of illness that are alleged to be related to pesticides, and to provide an overview to inform the pesticides approvals process. [94] The role of the HSE is further discussed in 5. 90-5. 91.

The PIAP costs between £30, 000 and £40, 000 per annum to run. Its primary role is not to assess causality in individual cases, [95] but to look for trends that might be indicative of associations between individual active ingredients and ill health effects, using a product-focused analysis.

2.81. Each year about 60 cases of complaints reach the PIAP system. A case will only reach PIAP after the HSE investigation and any prosecution is completed and the file closed, which on average takes six months. When PIAP receives papers from the HSE, the Panel assesses each individual case on the strength of the association between exposure and ill health, and each case is classified as being 'confirmed', 'likely', an 'open assessment' or having 'insufficient information'. The Panel meets every quarter and publishes an annual report of pesticide incidents and statistics. [96]

2.82. According to the HSE, as part of its inspection (2. 77) an Employment Medical Advisory Service (EMAS) physician will have investigated only 30% of the cases referred to PIAP. [97] PIAP seeks to obtain, with consent, additional medical records from general practice or hospital records for independent verification of symptoms. PIAP itself is not in a position to initiate further clinical examination or investigation of exposure (for example, biological or environmental monitoring) and this would in any event be unlikely to be useful taking into account the time-scales involved. [98] The delay before PIAP becomes involved means that it is unlikely that any of the pesticides implicated would still be present in environmental or biological media.

2.83. The Chair of PIAP, Dr Rawbone, told us that the Panel tends to pick up only acute effects from short-term exposure. [99] PIAP tends to receive only acute cases because chronic cases are not generally associated with 'incident investigation'. He stated that he has taken only one case of chronic illness to the Panel. [100] A problem with the role assigned to PIAP is that data on chronic health effects are strongly influenced by the effectiveness of incident reporting and the promptness with which incidents are investigated. [101]

2.84. PIAP does not feed information back to the original complainant as it primarily serves as a passive surveillance tool for risk assessment purposes. This is a cause of complaint for some of the residents and bystanders visited who had been referred to PIAP. It is unusual that the responsibility for a public health issue, such as bystander exposure to pesticides, rests with an executive agency such as the HSE that has primary responsibility for occupational health. The mission statement for the HSE is 'to protect people's health and safety by ensuring risks in the changing workplace are properly controlled', [102] and does not include a specific role for safeguarding public health. It is the role of the Department of Health (DH) to support the government in improving the health and wellbeing of people. [103] Indeed, evidence from residents and bystanders indicated that the current system, with PIAP as part of the HSE, leads to so much delay that the trail has gone cold by the time investigation by PIAP into allegations of ill health starts.

2.85. Given these problems, it is hardly surprising to hear residents and bystanders describing their experience of reporting arrangements under PIAP as highly unsatisfactory.

We believe that this system needs to be radically reformed, by the introduction of detailed clinical investigation, and extended to cover chronic cases. This is critical for an adequate understanding of the ill health effects attributed to pesticide exposure. To ascertain whether pesticides are indeed the cause of these adverse health effects it is important, not only that the numbers should be properly recorded through well-designed proactive surveillance methods, but also that a proactive investigative service should examine reported cases, where possible using modern laboratory methods such as imaging.

2.86. The DH's Health Protection Agency (HPA) remit is to protect the health and wellbeing of everyone in England and Wales. [104] In Scotland this responsibility lies within the parallel organisation Health Protection Scotland (HPS). The HPA has systems in place to deal with public health concerns about chemical contaminants, as well as having close links to the NHS, which is an important interface to the concerned public. However, currently the HPA has no remit to deal with public concern of pesticides, as when the HPA was set up the responsibility for pesticides was maintained within Defra, DH and HSE. [105]

Surveillance of Ill Health Effects from Pesticides

2.87. At present it is not possible to obtain precise figures for the number of people who claim that their health has been affected by pesticide spraying (1. 24). Considering that public concern over resident and bystander pesticide exposure has been acknowledged for some time, this situation is regrettable. While the PIAP system and other measures (such as the level of complaints received by manufacturers and the numbers quoted by lobby groups) might indicate that the number of residents and bystanders who complain of pesticide-related ill health is small, a definitive statement cannot be made without well-designed, national surveillance of the numbers of affected people and the range of health effects attributed by residents and bystanders to pesticide spraying.

2.88. A key difficulty for any surveillance and epidemiology is the paucity of information concerning what it is that people have been exposed to. Either this information is simply not known or the affected person cannot access the data readily. We discuss this issue further in 5. 75-5. 84.

2.89. There has been a number of attempts to assess the numbers of people who have been affected by pesticides. Appendix J discusses one of these studies, the Green Card Pilot scheme. It should be noted that this Green Card scheme and other schemes currently informing the regulators have all tended to focus only on the acute symptoms of exposure (2.83).

2.90. An inter-departmental working group chaired by the head of the HSE Health Directorate was set up in 1996 to review existing arrangement for monitoring the ill health effects of pesticides by government departments. The final report describes data sources for collecting information on the frequency and severity of acute pesticide poisoning in the UK. These included the PIAP scheme; the Home Accident Surveillance Scheme (HASS) reporting cases of ill health that occur in the home and leisure environment; Hospital Episode Statistics (HES) reporting ill health resulting in hospital admission; the National Poisons Information Service (NPIS) compiling information from physician enquiries about pesticide poisoning and ill health; and the Office for National Statistics providing mortality statistics from acute poisoning. [106] The working group identified a complete lack of information on cases dealt with by general practitioners, and lack of data relating to cases of chronic ill health. It recommended that HSE should, in consultation with others, explore the possibility of establishing a short-term scheme for the surveillance of pesticide-related illness brought to the attention of general practitioners, and that the extent to which there is a link between chronic ill health and pesticide exposure was best addressed through specific research projects. [107]

2.91. In its 1987 report the House of Commons Agriculture Select Committee stated that there was a clear need for a centralised system to co-ordinate all reports on pesticide poisoning. [108] In the time since the report was published the situation has not improved. [109]

2.92. There have been several more recent attempts to collate information on resident and bystander effects from pesticide spraying, involving surveillance through physicians.

The first of these is an HSE-commissioned project with the Institute for Environment and Health (IEH) that started in 2004, which is looking at the possibility of obtaining primary care information about pesticide-related ill health direct from general practitioners. This study aims to collect data on the frequency of pesticide cases presented to general practitioners and diagnosed with suspected adverse health effects. [110]

2.93. The second survey, which is more passive and is currently running on the TOXBASE database (2. 72), aims to collect data on the numbers of people being adversely affected by pesticides. When accessing the entry for one of a specified list of active substances, the user will read the following request:'The National Poisons Information Service is collaborating in an audit with the Pesticides Safety Directorate and the Health and Safety Executive to improve surveillance of pesticide exposures and incidents'and be asked to fill in a form online. [111]

2.94. We have been surprised by the inadequacy of the arrangements for monitoring human health effects in relation to pesticide spraying. We know that such monitoring studies would not be easy, [112] but we have also been told that they could be designed to address these issues. [113] Surveillance studies would clearly be made easier if information on the individual pesticides being used were made publicly available. This information would help to test the validity of the suggestion made by some residents and bystanders that hot spots of cancer incidents exist in particular areas where agricultural pesticides are widely used.

2.95. Surveillance systems are usually set up to collect data on specific diagnoses linked to environmental exposures (for example, radon and cancer). It is difficult to create a surveillance system that is able to collect data on clusters of disease that have nonspecific symptoms (i. e. syndromes) resulting in a clinical diagnosis that is far from clear.

This situation is made even more difficult in a situation where there is either no data or imprecise data on pesticide exposure. It is possible to set up surveillance systems for environmental and pesticide exposures with less than perfect exposure data, however, such systems have been set up in other countries. [114] Without a surveillance system in place to collect such data, it is not possible for regulators to be confident that chronic ill health does not result from pesticide exposure. Such information could be used to understand better any environmental trigger factors for chronic disorders. [115]

2.96. In evidence presented to us, the Chief Executive of the HPA was confident that, with adequate funding, a proactive system to deal with pesticide exposure in residents and bystanders could be set up within the current structure of the HPA in England and Wales. [116] Such a system could be based around a systematic approach that would need to be developed for general chemical exposure as well as pesticides. This would require the collection of information for surveillance, through a number of routes, including NHS Direct, primary care, and the National Poisons Information Service and its TOXBASE database (2.

72). Systems could be developed for alerting general practitioners and NHS Direct nurses to acute and chronic symptoms of pesticide poisoning. Toxicological baseline studies for pesticides both as a single chemical or combination of chemicals, and standard approaches to clinical investigations would be developed. Importantly, a telephone hotline would be needed, possibly through NHS Direct (a 24-hour nurse advice and information helpline that operates in England and Wales), so that if a member of the public was concerned that they had been exposed to a pesticide they could get immediate advice. The HPA Chief Executive estimated that such a system would cost between £5 and £10 million per annum to run. [117] In Scotland, parallel organisations to the HPA and NHS Direct could use the same model.

Recommendations

2.97. We are concerned that those individuals who claim that their ill health relates to pesticide exposure have not had adequate access to well informed medical advice.

Following our evidence session with the Chief Medical Officer (CMO), Sir Liam Donaldson, a working group has been set up by the Department of Health. With the Royal College of Practitioners providing the Chairman, its aim is to produce a module of toxicology that could be integrated into the curriculum for general practitioners. [118] We endorse the Department of Health's move to strengthen higher professional development in the field of toxicology within general practice. They should also ensure that professionals working in public health and specialised poisons centres have a clear awareness and understanding of how to investigate the chronic health problems related to pesticides by residents and bystanders.

2.98. Recognising the deficiencies in expertise in health professionals of how to diagnose and treat individuals who clam to have chronic health effects attributed to pesticide spraying: we recommend that the Royal Medical Colleges agree how patients with chronic symptoms associated with pesticide spraying should be investigated and treated and identify a clear referral pathway from primary care to an appropriately trained consultant specialist.

2.99. We are concerned that those individuals who had considered that their illness was related to pesticide spraying faced considerable difficulties in registering their concerns with a government agency that would lead to possible further investigation.

Following our evidence session with the Chief Medical Officer (CMO), he has proposed a number of measures to improve the standards of care for members of the public concerned that their health has been adversely affected by pesticides. [119] These measures include, precise case definition, development of a database recording incidents and complaints, and the need for thorough investigation of alleged cases of harm. The CMO also suggested that there should be a more focused approach to reporting of bystander exposure from all sources of available information. [120] We recommend that a new national reporting and monitoring mechanism for ill health associated with pesticide spraying should replace the Pesticides Incident Appraisal Panel (PIAP) and that this should fall within the remit of the HPA and related organisations in the devolved administrations.

2.100. The lack of information relating biomarkers to adverse health effects of passive exposure to pesticides has hampered progress. Building on our proposals in 2. 68: we recommend that newly acquired and standardised clinical, physiological and laboratory information from those who attribute their adverse health effect to passive pesticide exposure should form the basis for future biological monitoring. In addition to those exposed to agricultural spraying, we recommend the establishment of a system that places greater emphasis on surveillance for adverse health effects of pesticides.

CONCLUSIONS

2.101. In its 1990 report, the British Medical Association (BMA) identified a lack of knowledge about the frequency of acute pesticide poisonings, possibly due to under-reporting, and about the effects of pesticides in causing chronic ill health. The BMA stated that this lack of knowledge needed to be corrected, although it recognised information would be difficult to obtain, and that given the uncertainty, it would be difficult to make categorical statements on the degree of risk posed by pesticides, particularly to human health. [121]

2.102. The BMA emphasised that the lack of information on how to gain access to data was perhaps the single greatest deficiency identified with regard to pesticide safety, and stated that existing information needed to be brought into the public domain and greater efforts made to obtain more complete, comprehensible and validated toxicological, epidemiological and other scientific data on chemical pesticides and their effects on human health. [122] It is disappointing that some fifteen years later these criticisms remain largely unaddressed.

2.103. The Chief Medical Officer, Sir Liam Donaldson, stated in his evidence to the Commission[123] that he felt that there was a need to be able to do something for individuals who feel that they may have been harmed as a result of exposure to pesticides, either through clinical examination or by trying to establish some causal relationship between their exposure and individual outcome. A standard protocol for biological monitoring needs to be developed, but more importantly the clinical awareness of general practitioners and specialists needs to be improved in order to improve the investigation of people with chronic, ill-defined health effects who state that they have been exposed to pesticides[124] or other environmental influences. The CMO followed his evidence to the Commission with a letter sent to the Commission, that outlined the feasibility of some of the ideas discussed. [125] These points are outlined in 2. 97 and 2. 99.

2.104. Some affected residents and bystanders seem to have turned to more unconventional, less regulated alternative approaches to medicine (such as chelation therapy and provocation/neutralisation antigen vaccines), because of their frustration with, what they see as, a lack of engagement within the conventional medical profession and a sense that their strong concerns are not being taken seriously by those working in the NHS.

2.105. There has been almost no effort to develop a systematic approach to deal with the complaints from residents and bystanders of ill health that they have associated with pesticide spraying. This situation needs to be changed. From what we have heard, residents and bystanders feel that they have been let down by the medical profession and the various government agencies charged with protecting their health. We have tried to review the evidence afresh and to reconsider the hypothesis that their reported ill health may be linked to pesticide exposure. We are not persuaded that the evidence from individual cases is so weak as to rule out this possibility.

2.106. We were encouraged by the open position taken by the Chief Medical Officer in his evidence and expect that this, together with the fact that the Minister Alun Michael took the initiative in asking us to review the evidence, will mean that government will act quickly to address the urgent need for changes in this area, to address the concerns of residents and bystanders, and to take steps to clarify this issue of potential risk to public health.

2.107. People are exposed to pesticides through uses other than in agriculture, such as amenity and timber treatment and these people have reported similar health concerns to those discussed in this chapter. Although we have not studied these areas in detail, the measures we recommend on health issues and monitoring human health should be extended to cover these non-agricultural pesticides.

REFERENCES

[1] Commission Directive 91/414/EEC, concerning the placing of plant protection products on the market, *Official Journal* L230, 19. 8. 1991.

[2] House of Commons Agriculture Select Committee (1987). *Effects of pesticides on human health*:second special report:volume I, report and proceedings of the Committee, (The Body Report), HMSO.

[3] British Medical Association (1990). *Pesticides, Chemicals and Health.*

[4] Casida, J. E. , and Quistad, G. B. (1998). Golden age of insecticide research:Past, present, or future? *Annual Reviews of Entomolology,* 43:1–16.

[5] Harel, M. , Kryger, G. , Rosenberry, T. L. , Mallender, W. D. , Lewis, T. , Fletcher, R. J. , Guss, J. M. , Silman, I. , and Sussman, J. L. (2000). Three-dimensional structures of Drosophila melanogaster acetylcholinesterase and of its complexes with two potent inhibitors. *Protein Science* 9:1063-1072.

[6] Moretto, A. and Lotti, M. (1998). Poisoning by organophosphorus insecticides and sensory neuropathy. *Journal of Neurolology, Neurosurgery, and Psychiatry* 64:463–468.

[7] Ray, D. (1998). *Organophosphorus esters: An evaluation of chronic neurotoxic effects. A Report by the MRC Institute for Environment and Health.* Available on IEH website at http://www. le. ac. uk/ieh/pdf/sr5. pdf

[8] Wessler, I. , Kirkpatrick, C. J. and Racke, K. (1999). The cholinergic 'pitfall':acetylcholine, a universal cell molecule in biological systems, including humans. *Clinical and Experimental Pharmacology and Physiology.* 26:198-205.

[9] Wessler *et al* (1999).

[10] Kolaczinski, J. H. and Curtis, C. F. (2004). Chronic illness as a result of low level exposure to synthetic pyrethroid insecticides:a review of the debate. *Food and Chemical Toxicology.* 42, 697-706.

[11] Hennekens, C. and Burning, J. (1987). *Epidemiology in Medicine.* Mayrent, S. Ed. Little, Brown and Co.

[12] Evidence from Professor David Coggon, November 2004.

[13] Hennekens *et al* (1987).

[14] Ontario Group of Family Physicians. (2004). *Pesticides Literature Review.* Available on OGFP website at http://www. ocfp. on. ca/english/ocfp/communications/publications/default. asp?s=1#EnvironmentHealth

[15] Ministry of Agriculture, Fisheries and Food/Health and Safety Commission (1990) *The code of practice for the safe use of pesticides on farms and holdings* (Green Code), HMSO.

[16] Department for Environment, Food and Rural Affairs (Defra) (1996). Nitrogen Dioxide. Expert Panel on Air Quality Standards. HMSO.

[17] House of Commons Agriculture Select Committee (1987). *Effects of pesticides on human health*:second special report:volume I, report and proceedings of the Committee, (The Body Report), HMSO.

[18] Institute of Environment and Health (2004). *Pesticides and Parkinson's disease:a critical review.* A report by the MRC Institute for Environment and Health for the Department of the Environment, Food and Rural Affairs.

[19] Ibid.

[20] Seaton, A. (2005). *Genetic, environmental and occupational risk factors for parkinsonism and Parkinson's disease* (Geoparkinson). Final project report.

[21] Coghlan, A. (2005) *Exposure to pesticides can cause Parkinson's. New Scientist.*

[22] Department of Health, Prostate Cancer. http://www. advisorybodies. doh. gov. uk/coc/prostate. htm accessed February 2005

[23] Department of Health, Organochlorine Pesticides and Breast Cancer. http://www. advisorybodies. doh. gov. uk/pdfs/cc043. pdf accessed April 2005.

[24] Ontario College of Family Physicians (2004).

[25] Ibid.

[26] Ibid.

[27] The Ontario Group of Family Physicians (2005). *Chapter:Annual Report 2003-2004*, 17:1.

[28] Pesticides Safety Directorate, Advisory Committee on Pesticides Statement on the Pesticides Literature Review published by the Ontario College of Family Physicians. http://www. pesticides. gov. uk/acp. asp?id=1387 accessed March 2005

[29] Pesticides Safety Directorate, Advisory Committee on Pesticides minority statement on the Pesticides Literature Review published by the Ontario College of Family Physicians. http://www. pesticides. gov. uk/acp. asp?id=1550 accessed July 2005.

[30] *Report to the Royal Commission on Environmental Pollution on the Pesticides literature review* (Ontario College of Family Physicians) - Dr Michael Burr. January 2005.

[31] Petrie, K. J. and Wessely, S. (2002). Modern worries, new technology, and medicine. *British Medical Journal.* 324 (7339):690-1.

[32] Bornschein, S. , Hausteiner, C. . Zilker, T. , and Forstl, H. (2002). Psychiatric and somatic disorders and multiple chemical sensitivity (MCS) in 264 'environmental patients'. *Psychological Medicine.* 32 (8):1387-1394.

[33] A report of the CFS/ME working group:report to the chief medical officer of an independent working group. January 2002.

[34] Miljøministeriet, (2002). *Duft- og kemikalieoverfølsomhed- Multiple Chemical Sensitivity, MCS.* Miljøprojekt Nr. 741.

[35] Evidence from Dr Annette Scivier, February 2005.

[36] Wessely, S. , Hotopf, M. , and Sharpe, M. (1998). *Chronic Fatigue and its Syndromes.* Oxford; Oxford University Press.

[37] Ibid.

[38] A report of the CFS/ME working group (2002).
[39] Research Advisory Committee on Gulf War Veterans'Illnesses (2004). Scientific progress in understanding Gulf War Veterans'Illnesses:report and recommendations. http://www1. va. gov/rac-gwvi/docs/ReportandRecommendations_2004. pdf accessed March 2005.
[40] Evidence from Syngenta, March 2005.
[41] Maynard, R. (2001). Acceptable risk of environmental air pollution. *International Journal of Hygiene and Environmental Health*. 204: 203-206.
[42] Ibid.
[43] Evidence from Syngenta, March 2005.
[44] Ibid.
[45] Working group on the risk assessment of mixtures of Pesticides (WiGRAMP) WIG/2000/02 final working paper:cumulative and aggregate risk assessment of pesticides and other compounds.
[46] Evidence from Professor David Coggon, November 2004.
[47] Whitford, F. , Fuhremann, T. , Rao, K. , Arce, G. , Klaunig, J. (2003). *Pesticides toxicology:evaluating safety and risk.* Ed. Blessing, A. Purdue Pesticide Programs.
[48] Evidence from Syngenta, March 2005.
[49] Ibid.
[50] Ibid.
[51] Chao, C. , DeLaHunt, M. , Hu, S. , Close, K. , and Peterson, P. (1992). Immunologically mediated fatigue:a murine model. *Clinical Immunology and Immunopathology*. 64:(2) 161-165.
[52] Overstreet, D. and Djuric, V. (2001). A genetic rat model of cholinergic hypersensitivity:implications for chemical intolerance, chronic fatigue, and asthma. *Annals of the New York Academy of Sciences* 933: 92-102
[53] Evidence from the Pesticides Safety Directorate, November 2004.
[54] Adapted from Twenty-fourth Report, Figure 2-I, p. 20
[55] Evidence from IPCS/OECD, February 2005.
[56] Simmons, P. and Portier, C. (2002). Toxicogenomics:the new frontier in risk analysis. *Carcinogenesis*. 22:(6) 903-905
[57] IPCS/Toxicogenomics/03/1. (2003). *Toxicogenomics and the risk assessment of chemicals for the protection of human health*. Report prepared for the WHO/UNEP/ILO international programme on chemical safety
[58] Personal communication, Syngenta, February 2004.
[59] Department of Health, Committees on Toxicity, Mutagenicity and Carcinogenicity of Chemicals in Food, Consumer Products and the Environment. http://www. advisorybodies. doh. gov. uk/cotnonfood/ toxicogenomics. htm accessed June 2005.
[60] McLean, W. , Holme, A. , Janneh, O. , Southgate, A. , Howard, C. V. , and Reed M. (1998). The effect of benomyl on neurite outgrowth in mouse NB2A and human SH-SY5Y neuroblastoma cells *in vitro*. *NeuroToxicology* 19, :629-632.
[61] Copeland, K. , Checkoway, H. , McMichael, A. , and Holbrook, R. (1977). Bias due to the misclassification in the estimation of relative risk. *American Journal of Epidemiology*. 105:(5) 488-495.
[62] PSD/Defra Tender Specification Document CT3010. July 2003. *Biomonitoring for pesticide exposures.*

[63] Ibid.
[64] CDC. January 2003. Second National Report on Human Exposure to Environmental Chemicals.
[65] The German Environmental Specimen Bank (ESB). Federal Ministry for the Environment, Nature Conservation and Nuclear safety. November 2003.
[66] Ezzati-Rice T. M. , and Murphy R. S. (1995). Issues associated with the design of a national probability sample for human exposure assessment. *Enviromental Health Perspectives* 103 Supplement 3:55-9.
[67] CDC. January 2003. Second National Report on Human Exposure to Environmental Chemicals.
[68] Wright, D. and Welbourn, P. (2002). *Environmental Toxicology,* Cambridge Environmental Chemistry Series 11.
[69] Washington State, Department of Labor and Industries. http://lni. wa. gov/Safety/Topics/AtoZ/Cholinesterase/default. asp accessed 9th March 2005.
[70] Washington State, Department of Agriculture, http://agr. wa. gov/PestFert/Pesticides/WorkerProtection. htm accessed March 2005.
[71] Evidence from Dr Michael Yost, January 2005.
[72] WWF (2003). *Contamination.* Results of WWF's biomonitoring survey.
[73] Evidence from Prof Kevin Jones, December 2004.
[74] WWF (2003).
[75] Evidence from the Pesticides Safety Directorate, November 2004.
[76] John Cherrie. Biological Monitoring of Pesticide Exposures, research contract with Defra. PS2604.
[77] Ibid.
[78] Evidence from Department of Health, February 2005.
[79] Biobank, http://www. ukbiobank. ac. uk/about/overview. php accessed February 2005.
[80] University of York, Mother and Infant Research Unit, http://www. york. ac. uk/healthsciences/miru/primre-search. htm accessed February 2005.
[81] The German Environmental Specimen Bank (ESB). Federal Ministry for the Environment, Nature Conservation and Nuclear safety. November 2003.
[82] ALSPAC. http://www. ich. bris. ac. uk/alspacext/Default. html accessed June, 2001.
[83] Ibid.
[84] Grey, C. N. B. , Nieuwenhuijsen, M. J. , Golding, J. , and the ALSPAC Team. (2005). The use and disposal of household pesticides. *Environmental Research.* 97:109-115.
[85] Sherriff, A. , Farrow, A. , Golding, J. , and Henderson J. (2005). Frequent use of chemical household products is associated with persistent wheezing in pre-school age children. *Thorax.* 60 (1):45-9.
[86] British Medical Association (1990).
[87] The National Environmental Education and Training Foundation. March 2002. *Implementation Plan. National Strategies for Health Care Providers:Pesticides Initiative.*
[88] Evidence from Department of Health, February 2005.
[89] Department of Health (1996). *Pesticide Poisoning Notes for the Guidance of Medical Practitioners.*
[90] Evidence from Chief Medical Officer, Sir Liam Donaldson, March 2005.
[91] Evidence from the Health and Safety Executive, October 2004.

[92] Health and Safety Executive (1999a). *Reporting Incidents of Exposure to Pesticides and Veterinary Medicines.*
[93] Health and Safety Executive (1999a).
[94] Health and Safety Executive (1999a).
[95] Evidence from the Health and Safety Executive, October 2004.
[96] Health and Safety Executive (2004). *Pesticide Incidents Report.* Field Operations Directorate Investigations. 1 April 2003-31 March 2004.
[97] Evidence from the Health and Safety Executive, October 2004.
[98] Ibid.
[99] Ibid.
[100] Ibid.
[101] House of Commons Agriculture Select Committee (1987).
[102] Health and Safety Executive, http://www. hse. gov. uk/aboutus/index. htm accessed March 2005.
[103] Department of Health, http://www. dh. gov. uk/AboutUs/fs/en accessed March 2005.
[104] Health Protection Agency, http://www. hpa. org. uk/ accessed March 2005.
[105] Evidence from the Health Protection Agency, February 2005.
[106] Health and Safety Executive (1999b). *Monitoring by Government Departments of the ill-health effects of pesticides:a review of existing arrangements.*
[107] Ibid.
[108] House of Commons Agriculture Select Committee (1987).
[109] Ibid.
[110] Evidence from the Institute of Environment and Health, November 2005.
[111] Evidence from Health and Safety Executive, October 2004.
[112] Ibid.
[113] Evidence from the Health Protection Agency, February 2005.
[114] Ibid.
[115] Ibid.
[116] Ibid.
[117] Ibid.
[118] Chief Medical Officer, Sir Liam Donaldson, letter to the Commission dated 10 June 2005.
[119] Ibid.
[120] Ibid.
[121] British Medical Association (1990).
[122] Ibid.
[123] Evidence from the Chief Medical Officer, Sir Liam Donaldson, March 2005.
[124] Ibid.
[125] Chief Medical Officer, Sir Liam Donaldson, letter to the Commission dated 10 June 2005.

In: Crop Spraying and the Health of Residents...
Editor: The Royal Commission, pp. 47-65
ISBN: 1-60021-391-X

Chapter 3

EXPOSURE

3.1. To assess the risks to human health posed by spraying pesticides, it is necessary to determine both the intrinsic health hazards and the exposure that a resident or bystander is likely to experience. We discussed the potential hazards of pesticides to human health in chapter 2. Here we consider exposure, which will depend on the method of application of the pesticide to the crop, its subsequent redistribution and the way it comes into contact with the exposed individual.

3.2. We begin by considering current application practice and potential routes of exposure.

We focus on the production, transport and fate of pesticide droplets, all of which influence the extent of exposure. We then examine the UK approach to exposure assessment and discuss its validity for assessing exposure of residents and bystanders.

CURRENT APPLICATION PRACTICE

3.3. The most widespread method of applying pesticides in UK agriculture is by boom sprayer (figure 3-I). Spray is produced by forcing the spray liquid by hydraulic pressure through nozzles situated at intervals along the length of the boom. Different types of nozzles may be used, [1] and air assistance can be incorporated. For spraying established perennial crops, such as trees, bushes and tall plants, fan-assisted axial sprayers (figure 3-II) are generally used, so that the airflow directs the spray droplets onto the target crop.

Aerial spraying is rarely practised in the UK, but is still used widely in other parts of the world.

3.4. The frequency of spraying in any particular year can vary widely for both arable fields and orchards, depending on the crop grown and the requirements for pest, disease, weed control and also for cosmetic purposes. On average, arable crops are sprayed five times a year, varying from three times for spring barley to thirteen times for potatoes (1. 25). [4] Orchard crops are sprayed an average of thirteen times per year, varying from three times for cherries to eighteen times for some apples. [5] The more frequent spray applications in horticulture reflect the greater use of fungicides as opposed to insecticides.

3.5. Pesticide formulations are composed of active ingredients and co-formulants (appendix E and 1. 9). The co-formulants can significantly affect the nature of the droplets produced, for example through modifying the surface tension of the spray liquid. [6, 7, 8] In many cases the smell associated with a sprayed pesticide is attributable to co-formulants contained in the formulation (appendix E).

Figure 3-I. Boom spray and standard flat fan nozzle[2].

Figure 3-II. Axial fan-assisted spray[3].

ROUTES OF EXPOSURE

Spray Application and Drift

3.6. The extent of exposure is determined by the amounts of pesticide transported to the recipient and the mechanisms by which the pesticide is taken up. Pesticides from spray application may be transported as liquid droplets, vapour or particulates. Residents and bystanders may be exposed to any of these forms. Direct uptake may occur as a result of penetration through skin (dermal absorption) and eyes, or through inhalation. Indirect uptake may result from ingestion or contact with residues on clothing, surfaces, garden produce or dust. Following absorption through the skin, substances can either flow along the nervous system or through the blood stream. Residents and bystanders may be exposed at the time of application and possibly for a period thereafter until the pesticide degrades, becomes unavailable or is removed by other processes.

3.7. The activity and location of the receiving person is important when measuring and observing the effects of drift. Many exposure studies, and other studies looking at actual residue levels in the body and health effects, have looked at operators rather than bystanders. This is partly because operators are more easily identified and partly because of an assumption that they are likely to be exposed to higher doses of pesticide.

(a)

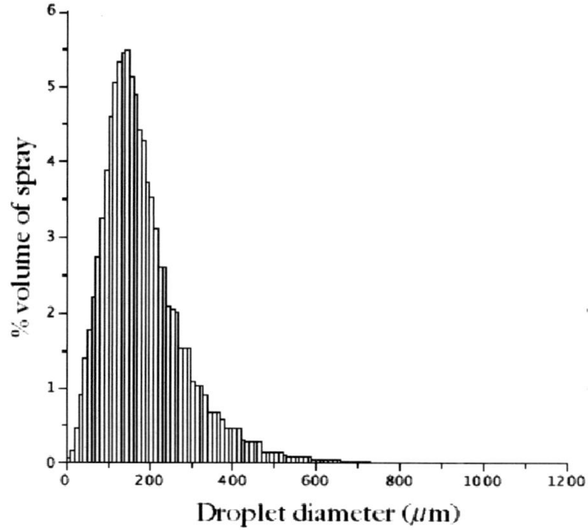

(b)

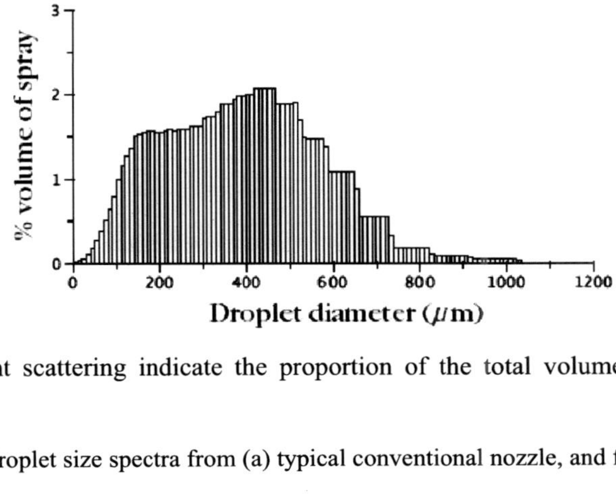

Measurements by light scattering indicate the proportion of the total volume that is in droplets of different sizes.

Figure 3-III. Pesticide droplet size spectra from (a) typical conventional nozzle, and from (b) LERAP three star rated nozzle[10].

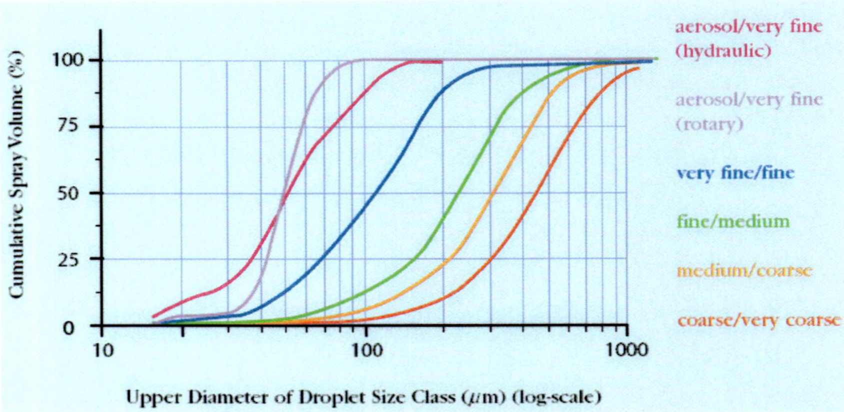

On this diagram any curve indicates, for a given diameter, the cumulative percentage of the total volume made up of all the droplets smaller than this value. Very fine nozzles correspond to a curve to the left of the first curve on the left, fine nozzles between the lefthand pair of curves, then medium, followed by coarse, and finally very coarse to the right of the last curve on the right. Note that the position of the rotary nozzle curve would depend on the speed of rotation and the flow rate of liquid on to the disc.

Figure 3-IV. Droplet size spectra thresholds from common nozzle designs[11].

Within the Green Code guidelines (chapter 5), the operator is recommended to wear personal protective equipment (PPE) (2. 23, 5. 01-5. 71) and to suitably shield the tractor cab from potential drift. A resident or bystander does not normally have access to such protective measures, and could be more exposed to the pesticide being applied (2. 23).

3.8. Atomisation of liquid pesticide formulation by conventional hydraulic pressure nozzles on a spray boom produces a wide spectrum of droplet sizes, with a predominant size range typically in the range 45–600 _m diameter (figure 3-IIIa). The British Crop Protection Council (BCPC) classifies the droplet spectra produced by different types of nozzle as very fine, fine, medium, coarse and very coarse, [9] by using a system of reference nozzles to overcome the problems of different measurement techniques. Corresponding droplet size distribution thresholds are shown in figure 3-IV.

3.9. The proportions of different size droplets emitted from a nozzle orifice depend largely on the design of the nozzle, [12] although other factors, notably pressure and spray volume rate[1] also affect the droplet size distribution. For example, droplet size decreases with increasing hydraulic pressure. The nature of the droplet spectrum is important because droplet size influences trajectories, deposition and efficacy against the target, as well as the risk and extent of drift.

3.10. Drift can be considered as the proportion of spray droplets that is still airborne at a specified distance downwind from the sprayer and is the main route of exposure to residents and bystanders. Figure 3-V illustrates the relationship between droplet size and distance transported before deposition. Larger droplets settle to the ground sooner than smaller droplets, which have a greater tendency to drift downwind. Coarser sprays, containing a higher proportion of larger droplets, thus produce the least drift. Finer sprays often give higher coverage and efficacy, but smaller droplets remain airborne longer and also allow

[1] Standard term used for the volume of spray applied per unit area (litres/hectare).

more opportunity for evaporation. Fine sprays, particularly smaller than about 100 μm diameter, are therefore particularly liable to drift (figure 3-V).

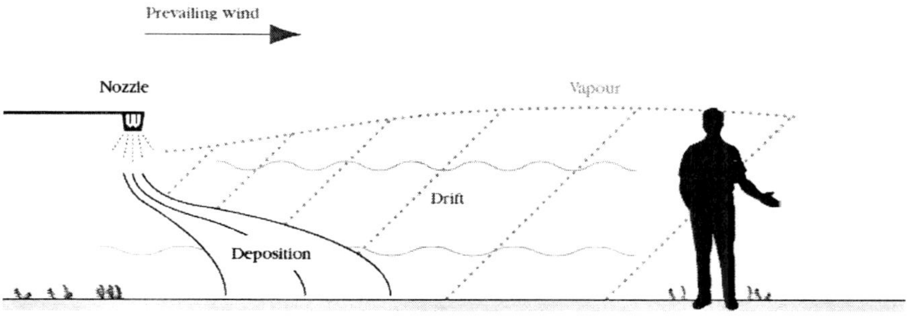

Figure 3-V. Relationship between droplet size and distance transported before deposition – schematic.

3.11. Figure 3-III and figure 3-V show that most agricultural sprays contain a proportion of droplets smaller than 100 μm, ranging from typically 5% in coarse or very coarse sprays to 20-50% for fine sprays and 50% for very fine sprays. Where there is a particular environmental risk, for example next to water bodies, the Green Code (chapter 5) recommends buffer zones (5. 85) or Local Environment Risk Assessment of Pesticides (LERAP) requirements (1. 29) to be followed. [13] The buffer zone size may be adjusted on the basis of compliance with the LERAP scheme, set out by the Pesticides Safety Directorate (PSD). [14] The LERAP scheme employs low drift equipment including special nozzles (figure 3-IIIb), which would normally be coarse or very coarse to reduce drift. Most recommended LERAP nozzles employ air induction, producing very large droplets that tend not to drift, and contain air bubbles, which make them less likely to bounce off foliage.

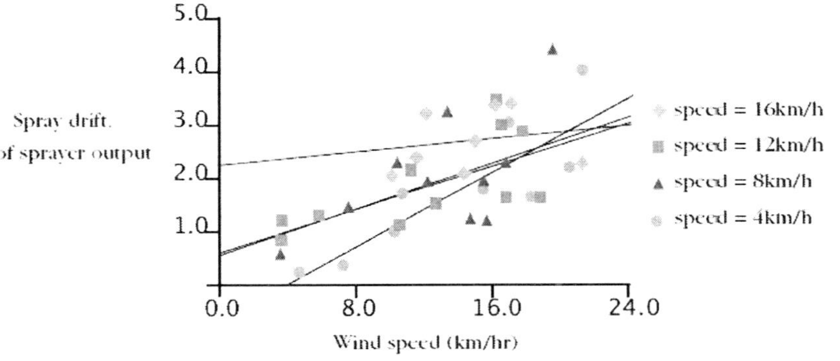

Figure 3-VI. Drift measured in field conditions at various wind speeds and for four tractor speeds[17].

3.12. Following droplet production, the behaviour of spray droplets depends on factors such as the initial pattern of release, tractor speed, the type of formulation, nature of co-formulants, local topography and meteorological conditions (particularly temperature and wind conditions). The airflow above crops or soil usually consists of a complex structure of turbulent eddies due to the drag from the rough surface and local temperature variation. The nature of this turbulence depends on many factors, including the vegetation, the topography, the wind strength and the stability of the atmosphere. Wind speed can significantly increase drift, [15, 16] (figure 3-VI) but other interacting factors such as wind direction, are important when considering the extent of resident and bystander exposure.

3.13. The net effect of the hydraulic pressure applied in the sprayer on drift is not straightforward, since drift is increased by a reduction in droplet velocity but decreased by an increase in droplet size. In addition, the relationship between spray volume rate and hydraulic pressure is not linear, so that adjustments in spray volume rate can have a disproportionate effect on the potential for drift; for example, doubling the spray volume rate for some given nozzles requires the pressure to be increased four fold. [18] However, farmers are advised only to adjust the pressure for small changes in flow rate (c5%) and to change the nozzle tip if they wish to apply a different spray volume rate.

3.14. The farmer can usually choose to reduce the volume of water used to dilute the pesticide to be applied, thus increasing the concentration of active ingredient in the spray. [19] The Green Code sets out situations when this is not acceptable. Although a spray volume rate of 200 litres/ha is normally advised in product instructions, it is not uncommon for operators to use volumes of 100 litres/ha or less, [20] as they can save money on transport costs by using the material in a more concentrated form. In these cases the concentration of active ingredient will obviously be higher than that assumed in the Pesticide Safety Directorate (PSD) standard exposure assessment (3. 28-3. 31).

Sometimes, however, farmers may use less active ingredient than the maximum, where they or their agronomist advisers consider that the lower dose is sufficient to deal with the problem.

3.15. Reducing application volumes can have significant advantages for the farmer. Firstly, it allows faster work rates. This can lead to improved timing of applications, which is one of the most important factors in the success of a pesticide application. With limited weather opportunities for spraying in the UK, being able to spray when the pest is most vulnerable is important in minimising overall pesticide use. Secondly, the higher concentration of pesticide at reduced volume may, for some products, improve efficacy.

Finally, the finer sprays associated with some nozzle types at low volumes may also give optimum efficacy, but the finer sprays also increase the risk and quantity of drift. A compromise may sometimes be reached, whereby farmers change the nozzle for the last downwind swath across a field so that the coarser spray is used at the edge of the field.

3.16. Further operational factors relevant to drift include tractor speed and boom height.

Increasing tractor speed generally increases the speed of the droplets with respect to the air and therefore the likelihood of them becoming airborne; this effect may be enhanced by wind, particularly if the tractor is driving into the wind. Figure 3-VI shows this, and also illustrates the wide variability, or scatter, of measurements made under field conditions. Boom height is more difficult to control at higher tractor speeds, especially on uneven ground. The importance of boom height has been indicated by wind tunnel experiments in which a change in height from 0. 5-0. 8 metres, led to a four-to five fold increase in the entrainment of particles in the air stream, with a corresponding increase in drift. [21] Optimum height of the nozzle on the boom is also a function of nozzle design:while a height of between 0. 35 and 0. 5 metres is appropriate for 110° flat fan nozzles, other nozzles may require different heights.

Evaporation and Volatilisation

3.17. Droplets emitted from spray equipment are also subject to evaporation. Evaporation of droplets is determined by meteorological conditions, especially temperature and humidity and is also strongly dependent on droplet size. In general, under the same conditions, the lifetime of airborne water droplets is proportional to the square of their diameters. Smaller droplets in a spray may therefore decrease in size significantly and even completely evaporate while larger droplets are still not much reduced in size.

3.18. The rate of evaporation of any component in a pesticide droplet depends on its vapour pressure. More volatile components – water and co-formulants – can evaporate to leave a very fine residual particle of relatively involatile active ingredient. Such very fine particles will drift further than the initial droplets. Exposure to the active ingredient as a vapour is more likely to result from volatilisation from these residual solid particles or from deposits on foliage and soil than from the spray droplets themselves. [22] In agricultural areas of inland California, evaporation from plant and soil surfaces has been found to be significant for some pesticides with moderate vapour pressure (10^{-1} to 10^{-2} Pa) and has been observed for some low vapour pressure pesticides (10^{-3} to 10^{-5} Pa). [23, 24] The significance of evaporation of low volatility pesticides is open to debate, but there is evidence that evaporation of such materials from plant and soil surfaces has led to elevated ambient air levels in some rural areas. [25] It is also possible that pesticide residues in soil could be made airborne and transported at a later time in dust particles.

3.19. Vapour uptake by vegetation may be particularly significant because molecules in air can penetrate by diffusion into every exposed surface. The significance of vapour effects then depends on the collection efficiency of the surfaces. In the exposed individual, active processes, such as respiration, may also enhance uptake. Figures are available for human respiration rates under various conditions from a wide range of sources; when combined with information on levels of substances present in air, these can be used to estimate exposure by this route.

3.20. The risk of exposure to vapour from deposits of pesticides after application has been a concern to residents and bystanders, but is difficult to quantify. Vapour effects outside the area of application involve a sequence of processes:evaporation from the target area, movement of the vapour downwind with associated turbulent dilution, and the presence of the resident or bystander in the non-target area. Each stage is complex and dependant on the particular local conditions, for example evaporation of deposited pesticide will be affected by sorption, uptake and decomposition on the receiving surfaces. Concentrations of vapour from deposits are therefore likely to decrease, sometimes rapidly, with time. The studies quoted in 3. 18 provide clear evidence of re-evaporation from deposits, but many such studies relate to the period immediately after application and are difficult to generalise. As with other aspects of exposure, given these difficulties, there is a need for more extensive monitoring of concentrations in air under field conditions in the UK (3. 53-3. 54)).

Particulates

3.21. The drift and interception of solid particles (for example, fine soil or dust from plant material at harvest) follows the same principles as for droplets, but shape can have an important additional influence on behaviour. In considering such possible exposure routes, account must be taken of degradation and further transport processes, for example within soil, which can substantially reduce the level of active ingredient.

Rates of degradation will depend on the chemical structure of the substance concerned, the ambient conditions and the nature of the receiving material, organism or individual.

3.22. It is clear that large droplets, slow tractor speed, optimal wind properties and some other factors that contribute to minimising drift, may at the same time compromise the efficacy of pesticide application. Reducing drift without taking efficacy and timing into account can result in higher levels of exposure for people and the environment.

Thus, it is very important to maintain an appropriate balance between efficacy and drift reduction.

Interception, Deposition and Retention of Spray Droplets

3.23. Whereas fluids, like air, flow around obstacles, entrained droplets or particles have higher inertia and may therefore impinge on the obstacle rather than passing around it with the fluid. Furthermore, turbulence in the flow may increase impaction of particles on the obstacle. Especially within the crop canopy, inertial deposition processes are complex and difficult to predict, but in general impaction increases with the ratio of droplet to object size and increases very strongly with air velocity.

Impact, retention and adhesion depend on the morphology of the receiving surface; for example, leaf hairs retain small droplets, whereas waxy surfaces can repel aqueous drops. Such processes can be manipulated by adjusting the physical properties of the spray liquid and the use of co-formulants.

3.24. Field boundaries or buffer zones can significantly reduce spray drift by capturing spray particles. [26, 27] For example, in one investigation, a two metre wide strip of grass one metre tall was found to reduce drift by up to 70% and a four metre wide strip by up to 85%. [28]

3.25. Vertical structures such as shrubs, trees, hedges and windbreaks (which may be more practicable than buffer strips in some horticultural situations) can intercept spray drift, with amounts intercepted increasing with proximity to the application area. However, such structures must be porous to be effective. For hedges, 40-50% porosity is optimum:a more dense hedge or wall lofts the drift cloud up and over the structure, resulting in higher deposition further downwind. [29, 30, 31, 32] The extent of drift is also strongly influenced by the stage of development of the crop, especially the amount of leaf cover; the risk of airborne drift may be as much as 50% greater early in the season when leaf cover is sparse compared with later when cover is complete. [33, 34]

3.26. The characteristics of interception, deposition and uptake of pesticide droplets and particles are of critical importance when measuring exposure. Spray drift can be expressed in different ways and there are measurement options appropriate for considering each exposure situation. For example, for deposition onto a watercourse or garden it is important to consider the deposition on a horizontal surface downwind from the sprayer as a proportion of the spray released per unit of the ground area. However, for inhalation it is important to consider the airborne concentration at head height.

The relationship between concentration of transported spray droplets and distance from the sprayer is not straightforward, as is apparent in figure 3-VII. At a distance of one and five metres from the sprayer the airborne concentration at adult height changes little. At a child's height close to the sprayer it is much higher, but drops off relatively rapidly with distance.

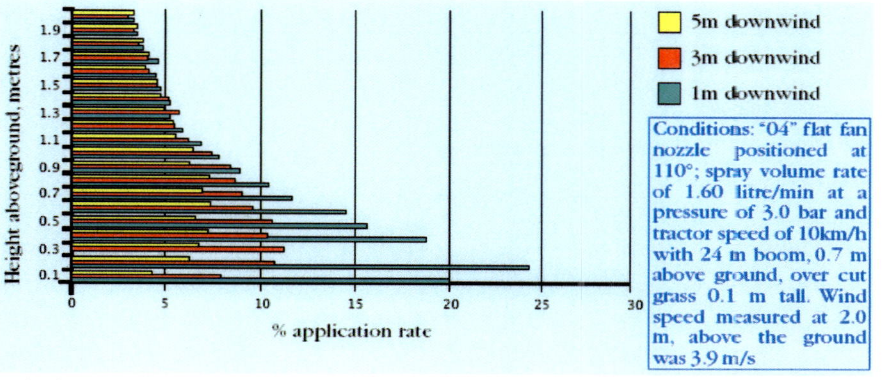

Note: the marked decrease with distance of the spray drift below a height of one metre because of the fall-out of larger drops. The overall reduction in spray drift with distance would be much larger with taller vegetation.

Figure 3-VII. Vertical profiles of airborne pesticide concentration with distance from application area in a field experiment with cut grass[35].

3.27. To sum up, it is clear that resident and bystander exposure to sprayed agri-chemicals depends on a range of factors and conditions, including spray composition and concentration, nozzle design and operation, meteorological conditions, field topography, crop foliage and buffer zones. These factors lead to significant variations in the likelihood of resident and bystander exposure and should be incorporated when predicting spray drift. We now examine the extent to which the prevailing approach to estimating resident and bystander exposure meets this requirement.

EXPOSURE CALCULATIONS MADE BY AND FOR THE PSD AND THE ACP

The Standard Exposure Assessment

3.28. The Pesticides Safety Directorate (PSD) aims to avoid risks to health from pesticide application by ensuring that exposure can never reach damaging levels. The PSD's standard assessment of exposure of bystanders to pesticides uses the formula in box 3A to calculate exposure from spray drift leading to uptake through the skin and by inhalation at the time of application and for a period of five minutes thereafter.

Application of a specific pesticide is considered to be acceptable if the predicted exposure never exceeds one hundredth of the Acceptable Operator Exposure Level (AOEL) determined from animal toxicity tests, with the uncertainty factor 1/100 included to allow for inter- and intra-species differences (2. 49-2. 50 and figure 2-I). As the AOEL is based on exposing animals to repeated doses of the active ingredient (2. 45), this approach should allow for repeated exposure due to repeated pesticide applications.

BOX 3A FORMULA USED BY THE PSD TO ASSESS EXPOSURE OF BYSTANDERS TO PESTICIDES
Exposure = [(PDE x SC x DA) + (PIE x SC x 100%)] / BW mg/kg body weight
Where:PDE = potential dermal exposure (ml) SC = concentration of active ingredient (mg/ml) DA = dermal absorption (%) PIE = potential inhalation exposure (ml) BW = body weight (kg)

3.29. The crucial parameters in the formula used by the PSD, describing potential dermal (PDE) and inhalation exposures (PIE), are not derived from modelling a range of experimental conditions, but are based on data from two sets of trials conducted in the 1980s by the then Ministry of Agriculture, Fisheries and Food (MAFF), with some support from industry. The first of these trials[36] for arable spraying was conducted in 1983 and the second[37] trial for orchard spraying was undertaken in 1987.

3.30. In the arable trials, a 12 metre boom was operated at a height of 0. 5 metre above the crop and at a tractor speed of 8km/h, spraying at right angles to the prevailing wind direction.

Four application rates and a standard nozzle were used. Measurements of the amounts of material absorbed onto skin or breathed in were made in specific locations during a number of single passes of the sprayer at a variety of wind speeds. The parameter values (PDE and PIE) used by the PSD are based on average measurements taken from the trials eight metres downwind from the sprayer and multiplied by a factor of three, which is assumed to account for the multiple passes that would be made in a real spraying situation. Calculating the contribution dermal exposure makes to overall exposure assumes 100% absorption of the PDE impacting on the skin of a person who is taken to be unclothed. The PIE is calculated in ml by multiplying the measured air concentration by the respiration rate for heavy work (20 litres/min) over a five-minute exposure period. This calculation of the contribution of inhalation to overall exposure assumes 100% absorption and retention of the inhaled spray.

3.31. The orchard parameter values were determined in a similar manner to those in the arable trials:spray drift levels were measured in the breathing zone and from that deposited on the clothing of bystanders located between 8 and 50 metres downwind for the 20-minute spraying period. The PDE and PIE for orchard spraying are then derived in the same way as for arable spraying and used to calculate an overall exposure level.

Further Exposure Calculations by PSD

3.32. Following concerns about the adequacy of the bystander exposure assessment, in particular those raised by rural resident and pesticide campaigner Georgina Downs, along with the presentation of a video showing pesticide exposure at an Advisory Committee on Pesticides (ACP) open meeting in July 2002, the ACP reviewed the exposure assessment approach in 2003. This review was based on two papers produced for the ACP by the PSD -

Bystander exposure assessment: ACP9(297/2003)[38] and *Bystander risk assessment, further exposure estimates:ACP20(301/2003)[39]* - which were tabled at the January and July 2003 meetings of the ACP. The papers focused on four aspects:exposure closer than eight metres to the sprayer, longer-term inhalation of vapour, exposure through dust, and the exposure of children (5. 26-5. 27).

3.33. In paper *ACP9(297/2003),* existing data from an experiment on the surface deposition of spray obtained for a study on watercourse protection[40] were used to infer potential dermal exposures from arable and orchard spraying at one and three metres respectively from the sprayer. For this purpose only, these were used as inputs to the formula in box 3A to estimate the exposure closer to the sprayer than the eight metres assumed in the standard exposure formula. No attempt was made to estimate droplet interception by an object or bystander in the path of the prevailing airflow. Also, no change was made in the standard estimate of potential inhalation exposure. [41] Estimates of dermal exposure calculated from depositional data following spraying at one and three metres were many times greater than the estimates calculated by the PSD standard formula on at least two occasions, though these were measured in wind speeds above those recommended in the Green Code (5. 74) as suitable for spraying.

3.34. PSD also collected new bystander exposure data from arable spray trials conducted by the Central Science Laboratory (CSL) in 2002 trials, similar to those from the 1980s MAFF trials. Dermal exposures were measured at one and five metres downwind from the sprayer. The new data confirmed that dermal exposure is greater closer to the sprayer. The highest PDE measured at one metre distance was seven times higher than the mean value used in the standard formula (box 3A), which is based on a distance of eight metres from the sprayer. Bystander exposure in these later trials was found to be reduced when air induction nozzles were used. The later trials also confirmed that wind speed is a major factor in determining the relationship between exposure and proximity of sprayer.

3.35. Paper *ACP9(297/2003)* also introduced estimates of exposure from long-term inhalation of pesticide vapour, a route not covered by the standard assessment formula in box 3A. The estimates were based on concentrations taken from an unpublished PhD thesis[42] and from German[43] and Californian studies. [44, 45, 46, 47, 48, 49] In paper *ACP20(301/2003)*, these long-term inhalation estimates were supplemented by more recent German data. [50] The first estimates of inhalation by adults and children for a range of pesticides were based on the Californian data[51] for chlorpyrifos, which is a relatively volatile pesticide. Where these initial estimates that were made using the PSD formula (box 3A) were large compared with the AOEL, more refined estimates were obtained using the concentration data available for a less volatile chemical with vapour pressure closer to the substance concerned. The final estimated exposure for several of the pesticides considered was comparable to, or greater than, the AOEL. However a number of these pesticides had already been withdrawn from use or were being reconsidered. In California and elsewhere in the USA, a considerable amount of aerial spraying still takes place in hot weather; the amount of active ingredient retained in the atmosphere as a result of evaporation is therefore probably greater than would be found under UK atmospheric conditions.

3.36. Estimates of bystander exposure to pesticides in soil dust were based on soil inhalation data for drivers of tractors with no cabs. In paper *ACP9(297/2003)* these were supplemented with estimates for dust at harvest from foliar sprays, this time based on dust levels experienced by drivers of uncabbed combine harvesters. [52] The estimates were described as speculative, but according to the paper, suggested that this route of exposure, taken alone, is not of concern.

3.37. Exposure of children to pesticides is a particular concern. As shown earlier (figure 3-VII), spray drift[53] levels can be much higher one metre above the ground than above one metre. Furthermore, children generally have higher respiration rates per body mass than adults and their detoxification and clearance systems are less developed, so that they are at a higher risk than adults to adverse effects from airborne toxins. [54] Small children may also be subject to an additional exposure route, by ingestion of residues through consumption of soil and other object-to-mouth contact. To incorporate this route, it is necessary to quantify transferable residues on soil and foliage, exposure duration, quantities of foliage and soil ingested, surface area of hands exposed and a saliva extraction factor.

3.38. Paper *ACP20(301/2003)* contained estimates of the pesticide exposure of children following spray drift deposition on the ground. [55] Total exposure was estimated as the sum of dermal, hand-to-mouth and object-to-mouth exposures. Each exposure route was calculated by inserting new parameters into separate formulae, similar to that used in the standard PSD exposure assessment (box 3A). Although these calculations considered dermal exposure close to the sprayer, they did not incorporate inhalation by children close to the sprayer. The ACP consideration of these papers is reflected in two documents headed 'Advice to Ministers'which are discussed in chapter 5 (5. 24-5. 32), and did not result in any changes in pesticides approved for use or policy.

DISCUSSION OF THE PSD EXPOSURE CALCULATIONS AND OF THE CONCLUSIONS DRAWN BY THE ACP

3.39. The standard exposure assessment used by the PSD represents a pragmatic approach to a complicated problem. Given the assumptions of an unclothed bystander, high breathing rate, proximity to the boom, the uncertainty factor of 100 that has been used in calculating the AOEL from animal toxicology data, and with some account being taken of repeated exposure, the assessment is probably conservative and protective in the majority of situations. However, the assessment only accounts for exposure through spray drift and essentially ignores the complex phenomena governing it.

There is no indication, for example, that the well-regarded research on spray drift that has taken place over many years at the Silsoe Research Institute, funded by the Department for Environment, Food and Rural Affairs (Defra), has been used to evaluate the validity of the approach used and assumptions made, or to develop the approach further. As a result, the PSD approach, defined by the standard formula (box 3A), suffers from the following serious shortcomings:

- it is based on a gross simplification of real exposure and it does not address all of the possible exposure routes. It is based on single-valued parameters from a limited set of experiments and does not attempt to represent the physical mechanisms that determine the extent of exposure and takes little account of the variability of dispersion conditions encountered in practice;

- it explicitly takes into account only exposure by spray drift at the time of spraying by bystanders in the narrow sense, as defined in 1. 19, (though the additional papers, *ACP9(297/2003)* and *ACP20(301/2003)*, contained preliminary estimates of some other exposures to residents and others including inhalation of vapours and ingestion of soil);
- it is based on very limited data obtained in the 1980s, since when, farm practices have changed significantly; for example tractors are capable of significantly faster speeds, boom sizes have increased, formulations have changed and the concentrations and flow rates used are often different;
- it assumes good operational practice; for example spraying only in recommended weather conditions (although high wind speeds were recorded during some trials) and at moderate vehicle speeds;
- average values from the 1980s trials were used to estimate the PDE and PIE, with no account taken of the large variability due to turbulence in the atmospheric boundary or of the dependence of drift on wind speed (3. 12, figure 3-VI);
- it does not take into account the important vertical variation in the profiles of spray drift (figure 3-VII) and their change with distance and wind speed (figure 3-VI);
- though a separate risk assessment is undertaken for re-entry to the sprayed field and subsequent contact with treated foliage, [56] which can be extended to individuals walking within a recently sprayed field, the standard formula does not include this route of exposure or risk assessment; and
- no account is taken of the possible importance for airflows and therefore spray drift, of topographic features such as sloping or irregular ground and also the effect they have on boom height, particularly if the tractor is driving at higher speeds.

3.40. Overall, we believe the PSD approach, even after its re-evaluation in 2003 which resulted in no change to the formula, is best described as yielding an approximate exposure estimate based on data which happened to be available, rather than a scientific prediction validated systematically by specifically designed experiments. Full details of the experimental conditions for the original 1983 and 1987 trials are not given and the approach has not been validated by the kind of peer review necessary for publication in the open scientific literature.

3.41. The results in papers *ACP9*(297/2003) and *ACP20*(301/2003) indicated a number of cases in which exposure could be comparable to or greater than the AOEL. But, to describe them as 'worst-case'estimates, as in the papers, is misleading. They are based on choices for the numerous parameters that are sometimes the median of the possible values, and other times the 75th or 90th percentile. Even at the 90th percentile level, more extreme values occur on 10% of occasions. Since the risk assessment considers a reference dose which is seen as sufficient to protect in case of repeated exposure, it might be suggested that extreme values are not relevant as they are likely to represent unusual events and would not reflect the dose to which someone would be repeatedly exposed.

3.42. A truly 'worst-case' analysis would consider the upper limit of the range of possible exposures and the probability of occurrence based on a statistical analysis of the available measurements. Furthermore, the sharp increase in spray drift below one metre in height and close to the sprayer seen in figure 3-VII could mean higher direct dermal and inhalation exposure for children, which appears to be ignored in the standard PSD approach. Given these points and those listed in 3. 39 over the completeness of the calculations, we believe that the categorical statement in the Discussion section of paper *ACP20(301/2003)* that, 'For products applied as sprays, these examples demonstrate that the current approach is protective of longer-term bystander exposure', does not reflect the uncertainties in the approach.

3.43. The ACP in its July 2003 discussion concluded in the paper: *Final Minutes of the 301st Meeting of the Advisory Committee on Pesticides* held on 10 July 2003:

3.1.3. Members felt that the paper provided a good review of the information available, and that the models used were appropriate and could be identified as worst-case scenarios.

3.1.4. A range of issues arising from the paper were discussed by members. It was agreed that the approach currently used to assess bystander risks is generally protective with the possible exception of soil fumigants. Further data were identified as necessary to complete the assessment for dithianon and trifluralin. '

3.44. We consider that the present approach may be conservative and protective in its treatment of targets, but in view of the absence of any attempt to model the complexity of bystander exposure and the probability of extreme values, we cannot agree that this has been conclusively or transparently demonstrated for the exposure process.

We cannot therefore support the ACP's unequivocal conclusion above.

THE GREEN CODE

3.45. As will be discussed in chapter 5, the Green Code[57] provides practical guidance on how to use pesticides safely under the existing legislation. It states that 'users of pesticides must ensure that all reasonable precautions are taken to prevent spray drift'. It also recognises that spray drift is affected by wind velocity, stability of the atmosphere, nozzle and pressure choice, vehicle speed, boom height, and equipment maintenance and settings, and advises that all of these should be considered and pesticide labels adhered to. It also advises '... considering not treating the boom-width or part of the boom-width closest to the boundary'. No mention is made of the possible influence of sloping ground and other topographic features.

3.46. The impact of weather conditions is highlighted, and the Green Code includes a table of wind speeds and the associated suitability of spraying. Spraying is inadvisable if the wind speed at boom height is Force 4 (9. 6-14. 5 km/h) or greater; however, no guidance is given on what to do if it is thought to be essential to spray crops during a period when the weather remains unsuitable. The best conditions for spraying are said to be Force 2 (3. 2-6. 5 km/h) blowing away from neighbours' land. There are warnings that little or no wind might be unsuitable because of very stable atmospheric conditions or the possible onset of convective conditions with gusty winds. There is also a warning that high temperatures with low humidity lead to increased evaporation of droplets and therefore decreased size and enhanced drift.

3.47. No firm advice on limits on the speed of the spraying vehicle is given. In practice, the CSL 2002 *Survey of Current Sprayer Practices in the UK*[58] indicates that 5-8% of tractor mounted and trailed sprayers use a speed greater than 16km/h, double that used in the 1980s trials on which the PSD exposure assessment formula is based. No guidelines are given for boom height, which is often determined by nozzle design (3. 15), and is frequently above the low value of 0. 5 metre used in the 1980s trials.

3.48. Users of sprayers are advised that the Met Office will provide useful information on wind speed, and that on reaching the field they should check the wind speed and direction using a simple wind meter or by watching the movement of trees and clouds. According to the CSL 2002 survey[58] on actual spraying practice, about 53% of spray operators used a meter to determine wind speed suitability, 36% looked for visual clues which could only be a guide for conditions near the boom, 3% used only a weather forecast, and 8% did nothing.

3.49. The Green Code raises the necessary issues for controlling spray drift. However, much of the advice is vague and adherence to the Code is not a statutory duty (chapter 5). Even if followed, the advice would not necessarily lead to practice similar to that used in the trials and for which the PSD exposure formula is likely to be most applicable. The analysis in this chapter strongly suggests that changes should be made to the Green Code and its application in order to reduce the probability of resident and bystander exposure to pesticides. Recommendations for such changes are made in chapter 5.

Conclusions

3.50. We have serious concerns about the current method of assessing resident and bystander exposure to pesticides. Although uncertainty factors are built into the AOEL, they are there to cover issues related to toxicology and do not address the variability of exposure or the uncertainties in exposure assessment.

3.51. We have also been disappointed to find that the current approach has not been rigorously evaluated under field conditions and has largely been assessed in relation to experiments done on a limited scale over twenty years ago and reassessed on the basis of other data often collected for different purposes in Germany and the US. The current approach does not take account of the range of possible factors that influence the probability of high exposures. It is important that new field testing is performed and that it should take account of variations in conditions such as boom height, wind speed and tractor speed so as to provide a sensitivity analysis. This will enable legislators and operators to have a better understanding of which variables are really important and how significant an impact they can make.

3.52. Neither does the current approach reflect the probabilistic nature of exposure of residents or bystanders to pesticides. In reality, the extent to which there is a problem will be determined by the probability of high resident and bystander exposure to a particular pesticide, together with the probability that this resident or bystander is sufficiently sensitive to that pesticide to be affected by such an exposure.

3.53. We recommend that the current approach for assessing resident and bystander exposure should, with some urgency, be replaced by a computational model which is probabilistic, looks at a wider range of possible exposure routes and more robustly reflects worst-case outcomes. The model should be rigorously validated by wind tunnel and field tests designed for the purpose, including non-standard conditions to test the sensitivity of the model predictions.

As a first step, whoever takes ownership of the creation of relevant data should undertake a thorough review of the relevant experimental work that already exists.

3.54. We recommend that validation of the model should be supported by monitoring of representative pesticides in a range of field situations. Such monitoring should include measurements of concentrations in air for extended periods after spraying to contribute to better understanding of possible exposure other than through immediate spray drift.

3.55. We recommend that both the model and the test data be peer reviewed and published in the open scientific literature so that the basis for the predictions is transparent and can be evaluated. Without this, there can be no expectation of confidence in the predictions.

3.56. In the short term, whilst the new probabilistic model is being developed and introduced, we recommend that all actual spraying practice be brought into line with the aspirations of the Green Code recommendations (chapter 5) including giving proper regard to the importance of optimal timing of the application and therefore efficacy of the pesticide. This will require appropriate monitoring arrangements and sanctions for non-compliance.

These short-term practical measures must contain provisions for recording relevant data.

REFERENCES

[1] Matthews, G. (2000). *Pesticide Application Methods.* 3rd Edition. Blackwell Scientific, Oxford, U. K.

[2] Kindly supplied by Silsoe Research Institute.

[3] Ibid.

[4] Garthwaite, D. G. , Thomas, A. R. , Dawson, A. , and Stoddart, H. (2002). *Pesticides Usage Survey Report: Arable crops in Great Britain 2002.* CSL. Available at: http:// www. csl. gov. uk/ science/ organ/ pvm/ puskm/ arable 2002. pdf.

[5] Garthwaite, D. G. , Thomas, A. R. , Dawson, A. , and Stoddart, H. (2002). *Pesticides Usage Survey Report:Orchard crops in Great Britain 2002.* CSL. Available at:http://www. csl. gov. uk/science/ organ/ pvm/ puskm/ orchards 2000. pdf.

[6] Miller, P. C. H. , Butler Ellis, M. C. (2000). *Effects of formulation on spray nozzle performance for applications from ground-based boom sprayers.* Crop Protection, 19:609-615.

[7] Butler Ellis, M. C. , Tuck, C. R. , Miller, P. C. H. (2001). *How surface tension of surfactant solutions influences the characteristics of sprays produced by hydraulic nozzles used for pesticide application.* Colloids and Interfaces:Physicochemical and Engineering Aspects, 180 (3) 267-276.

[8] Butler Ellis, M. C. , Bradley, A. (2002). *The influence of formulation on spray drift.* Aspects of Applied Biology, 66:251-258.

[9] Doble, S. J. , Matthews, G. A. , Rutherford, L. , Southcombe, E. S. E. (1985). *A system for classifying hydraulic and other atomisers into categories of spray quality.* Proceedings British Crop Protection Conference – Weeds, 3: 1125-1133.

[10] Miller *et al* (2000).

[11] Adapted from Matthews, G. A. and Bateman, R. P. (2004). *Classification criteria for fog and mist application of pesticides.* Aspects of Applied Biology, 71:55-60.

[12] Matthews, G. (2000).
[13] Defra, HSC and NAWAD (2004). *The draft code of practice for the safe use of plant protection products* (Green Code).
[14] The Pesticides Safety Directorate. *Local Environment Risk Assessment for Pesticides.* Available at: http://www. pesticides. gov. uk/safe_use. asp.
[15] Miller, P. C. H. , Smith R. W. (1997). *The effect of forward speed on the drift from boom sprayers.* Proceedings Brighton Crop Protection Conference – Weeds, 399-407.
[16] Thistle, H. W. (2004). *Meteorological concepts in the drift of pesticides.* Proceedings International conference on pesticide application for drift management, October, Washington State.
[17] Kindly supplied by Silsoe Research Institute.
[18] Miller *et al* (2000).
[19] Evidence from Silsoe Research Institute, January 2005.
[20] Evidence from the Pesticides Safety Directorate, April 2005.
[21] Evidence from Silsoe Research Institute, January 2005.
[22] Ibid.
[23] Ramaprasad, J. , Tsai, M. , Elgethum, K. , Herbert, V. R. , Felsot, A. , Yost, M. G. , Fenske, R. A. (2004). *The Washington aerial spray drift study:assessment of off-target organophosphorous insectide atmospheric movement by plant surface volatilisation.* Atmospheric Environment, 38:5703-5713.
[24] Woodrow, J. E. , Seiber, J. N. (1997) *Correlation techniques for estimating pesticide volatilization flux and downwind concentrations.* Environmental Science Technology, 31:523-529.
[25] Lee, S. , McLaughlin, R. , Harnly, M. , Gunier, R. , Kreutzer, R. (2002). *Community exposures to airborne agricultural pesticides in California:ranking of inhalation risks.* California Department of Health Services, Environmental Health Investigations Branch. Environmental Health Perspectives. 110:12.
[26] Evidence from Silsoe Research Institute, January 2005.
[27] Wolf, T. M. , Cessna, A. J. (2004). *Protecting aquatic and riparian areas from pesticide drift.* Proceedings International conference on pesticide application for drift management, October, Washington State University.
[28] Miller, P. C. H. , Lane, A. G. (1999). *Relationship between spray characteristics and drift risk into field boundaries of different structure.* Aspects of Applied Biology, 54. Field margins and buffer zones:ecology, management and policy, 45-51.
[29] Evidence from Silsoe Research Institute, January 2005.
[30] Brown, R. B. , Stephenson, G. R. (2001). *Spray buffer zone requirements with low-drift nozzles.* Guelph, Ontario:Ontario Ministry of Agriculture, Food and Rural Affairs, Report SR9055.
[31] Carter, M. H. , Brown, R. B. , Bennet, K. A. , Leunissen, M. , Kallidumbil, V. S. , Stephenson, G. R. (2000). *Methods for reducing buffer zone requirements for pesticide spraying adjacent to wetland environments.* Proceedings Sainte-Anne-de-Bellevue, Quebec 2000 National Meeting – Weeds. Available at:www. cwss-scm. ca/pdf/ ECW2000Proceedings. pdf.
[32] Davis, B. N. K. , Lakhani, K. H. , Yates, T. J. , Frost, A. J. , Plant, R. A. (1993). *Insecticide drift from ground-based, hydraulic spraying of peas and Brussels*

sprouts:bioassays for determining buffer zones. Agricultural Ecosystems Environment. 43:93-108.

[33] Syngenta (2005). Environmental Stewardship:Drift management. *Key practices to minimise drift - a checklist of possibilities.* Accessed at:www. syngentacropprotection-us. com/enviro/driftmanagement/ index. asp?nav=key_practices

[34] Evidence from Silsoe Research Institute, January 2005.

[35] Defra final report PS 2001. *Assessing the validity of current approaches for calculating off-crop exposure to pesticides.*

[36] Lloyd, G. A. , Bell, G. J. (1983). *Hydraulic nozzles:comparative spraying drift study.* MAFF.

[37] Lloyd, G. A. , Bell, G. J. , Samuels, S. W. , Cross, J. V. , Berries, A. M. (1987). *Orchard sprayers:Comparative operator exposure and spray drift study.* MAFF.

[38] ACP (2003). ACP9(297/2003) *Bystander exposure assessment.* ACP, U. K.

[39] ACP (2003). ACP20(301/2003) *Bystander risk assessment – further exposure estimates.* ACP, U. K.

[40] Rautmann, D. , Streloke, M. , Winkler, R. (2001). *New basic drift values in the authorisation procedure for plant protection products.* In Fortser, R. , Streloke, M. , Workshop on risk assessment and risk mitigation measures in the context of the authorisation of plant protection products (WORMM). Mitt. Biol. Bundesanst. Land-Forstwirtsch. Berlin-Dahlem, Heft 381.

[41] Glass, C. R. , Mathers, J. J. , Harrington, P. , Gilbert, A. J. , Smith, S. (2002). Project number:PA1728. Report number: 1728/02/02. Interim report field validation of LERAP (and assessment of bystander contamination).

[42] Turnbull, A. B. (1995). *An assessment of the fate and behaviour of selected pesticides in Rural England.* PhD Thesis, Department of Chemistry, University of Birmingham.

[43] Siebers, J. V. , Wittich, K. P. , Böttcher, S. , and Haenel, H. D. , (1999). *Untersuchungen zur Luftbelastung durch Pflanzenschutzmittel in der Nähe behandelter Flächen, Nachrichtenbl. Deut. Pflanzenschutzd,* 51 (9), 221-226.

[44] California Environmental Protection Agency, (1996). Report EH 95-10: *Summary of Assembly Bill 1807/ 3219 Pesticide Air Monitoring Results Conducted by the California Air Resources Board 1986 to 1995.*

[45] California Environmental Protection Agency, Air Resources Board, (2000). Report for the application and ambient air monitoring for atrazine.

[46] California Environmental Protection Agency, Air Resources Board, (1998). Report for the application and ambient air monitoring for chlorpyrifos (and the Oxon Analogue) in Tulare county During Spring/Summer 1996.

[47] California Environmental Protection Agency, Air Resources Board, (1999). Report for the application and ambient air monitoring for linuron in Kern county.

[48] California Environmental Protection Agency, Air Resources Board, (1999). Report for the application and ambient air monitoring for malathion in Imperial county.

[49] California Environmental Protection Agency, Air Resources Board, (1999). Report for the application (Tulare county) and ambient (Fresno county) air monitoring for simazine.

[50] Siebers, J. , Binner, R. , and Wittich, K. P. (2003). *Investigation on downwind short-range transport of pesticides after application in agricultural crops,* Chemosphere, 51:397-407.

[51] California Environmental Protection Agency, Air Resources Board. (1998).

[52] HSE (2002). Agricultural Information Sheet No 3 (rev). Controlling Grain Dust on Farms. HSE. U. K.

[53] Evidence from Silsoe Research Institute, January 2005.

[54] Ramaprasad, J. *et al* (2004).

[55] Rautmann, D. *et al* (2001).

[56] Evidence from the Pesticides Safety Directorate, April 2005.

[57] Defra, HSC and NAWAD (2004).

[58] Garthwaite, D. (2002). *Pesticide survey report:A survey of current farm sprayer practices in the UK.* Accessed at:http://www. csl. gov. uk/science/organ/pvm/puskm/cpa-finalpublication. pdf. CSL.

[59] *Ibid.*

In: Crop Spraying and the Health of Residents...
Editor: The Royal Commission, pp. 67-75 ISBN: 1-60021-391-X

Chapter 4

LEGAL LIABILITY FOR PESTICIDE EXPOSURE

4.1. In this chapter we consider possible areas of legal liability in cases of resident and bystander exposure, the difficulties of taking successful action and any need for reform.[1] We conclude that there are important threshold requirements that make successful action difficult, if not impossible.

TORTIOUS LIABILITY

4.2. The relevant legal principles governing the potential liability arising from the use of agricultural pesticides are primarily principles of tort law. Torts are civil wrongs that give rise to legal liability. These wrongs result in an injury or harm constituting the basis for a claim by the injured party. While some torts are crimes, the primary purpose of tort law is to permit the injured person to recover damages from the person imposing the harm. Tort law therefore primarily aims at compensation rather than prevention, although the person harmed may seek an injunction to prevent harmful conduct taking place or continuing, and the standards established by the courts will deter others from committing similar acts in the future. Among the types of damages the injured party may recover are compensation for loss of earnings or pain and suffering, or reasonable medical expenses. [1] Like all the common (judge-made) law, the scope of tort law is not static. It develops on a case-by-case basis, absorbing new evidence, modern standards and social, scientific and other factors.

4.3. Potential liability exists under several different heads of tort law. These are set out in more detail in a report written for the Commission by John Murphy (University of Manchester, School of Law) which can be found on our website. The following sections outline in brief the relevant heads of liability and the principles that allow, or stand in the way of successful private actions.

[1] This chapter reports the law of England, Wales and Northern Ireland. We do not believe that the law of Scotland is significantly different, but we have not been able to carry out a separate study.

Private Nuisance

4.4. A private nuisance is usually defined as any activity or state of affairs causing a substantial and unreasonable interference with a person's land or his or her use or enjoyment of that land. The kinds of activities that may give rise to a successful action in nuisance commonly include the emission of noxious fumes and smoke, as well as the generation of violent vibrations or noises that disturb the peaceful enjoyment of one's home.

Nuisance is grounded in 'strict liability', that is, liability irrespective of fault or intention to harm.

4.5. There are, however, important threshold requirements of nuisance law that make a successful action difficult in the case of pesticide spraying, namely:

- the claimant must be able to establish that he or she has suffered a 'substantial interference'. In this regard, it has long been established[2] that one cannot rely on one's own hypersensitivity in order to establish the substantial nature of interference;
- the claimant must have a proprietary interest in the land affected. Bystanders who do not own or occupy the land do not possess such an interest;
- strictly, there is no right to sue for personal injuries in the law of nuisance; it is a tort against land. It follows from the fact that nuisance is a tort against land that there is no right to sue for personal injuries. However, personal injuries might be recoverable where damage to the land was said to exist in the inability of those in occupation to enjoy it fully;
- the location of the claimant's premises will determine the level and kind of 'give-and-take'expected. Thus, where land is in an area devoted to farming, the occupier would be expected to bear the consequences of normal farming use;
- the fact that the defendant's activity is a 'reasonable use of land'will tend to protect him from liability. Thus compliance with the Control of Pesticides Regulations 1986, other regulatory instruments and the informal 'Green Code' will go a long way towards demonstrating reasonableness of use. Similarly, a failure to meet the conditions of the regulations may be indicative of unreasonable use; and
- an activity that is ostensibly capable of grounding a nuisance action may in fact become legitimate following 20 years overt, continuous and undisturbed use. The right to do something that would otherwise constitute private nuisance may be acquired by what is known as 'an easement by prescription;

Public Nuisance

4.6. A public nuisance – which is normally punishable as a crime – may nonetheless found an action in tort where a private citizen can show that he or she has suffered 'special damage' beyond that suffered by a sufficiently broad class of Her Majesty's subjects who are also affected by the defendant's activities.

4.7. If crop spray can be shown to affect a broad class of persons, but only cause physical illness in a few, an action on the basis of public nuisance may be possible, especially where non-compliance with the various regulations could also be shown.

The Rule in Rylands v Fletcher

4.8. The tort known as the rule in *Rylands v Fletcher* provides that:

The person who for his own purposes brings on his lands and collects and keeps there anything likely to do mischief if it escapes, must keep it in at his peril, and if he does not do so, is *prima facie* answerable for all the damage which is the natural consequence of its escape.

4.9. Like private nuisance, this tort does not require the claimant to show fault on the part of the defendant. Furthermore, the claimant need not show that the substance which escapes from the defendant's land is dangerous *per se*. The *Rylands* case itself concerned the escape of water from a reservoir.

4.10. On the other hand, there are also certain peculiarities concerning this tort that make a successful action far from straightforward, not the least of which is that personal injury recovery is contentious. There is a school of thought that suggests that personal injuries should not be recoverable under the rule in *Rylands v Fletcher*.

There is no authoritative statement of law on the matter.

Occupiers' Liability

4.11. Generally, anyone who lawfully visits property affected by the use of pesticides may well have a right of action against the occupier of those premises. The law distinguishes between liability to 'visitors'and 'non-visitors'.

4.12. By virtue of the Occupiers'Liability Act 1957, occupiers are under a duty to 'take such care as in all the circumstances of the case is reasonable to see that the visitor will be reasonably safe in using the premises for the purposes for which he is invited or permitted by the occupier to be there'. [3] Farm hands and other employees would be afforded the same duty by virtue of an implied contract term to this effect imposed by section 5 of the 1957 Act.

4.13. Those using a public right of way that crosses an occupier's land are not, however, to be treated as visitors for the purposes of the legislation. [4] Neither are those exercising a right of access to open land for recreational purposes afforded by the Countryside and Rights of Way Act 2000. [5] In both cases, the potential claimants fall into the category of 'non-visitors'who must look instead to the Occupiers'Liability Act 1984.

4.14. Under section 1(3) of the 1984 Act, non-visitors will only be owed a duty of care if (a) the occupier was aware of the danger or had reasonable grounds to believe it existed; (b) the occupier knew, or had reasonable grounds to know, that the uninvited entrant either was in, or might come into, the vicinity of the danger; and (c) the risk of injury to an uninvited entrant resulting from that danger was one against which, in all the circumstances of the case, the occupier might reasonably be expected to offer the uninvited entrant some protection.

4.15. It would be at least arguable that the lawful, closely regulated use of pesticides could be invoked to defeat any claims brought by, say, countryside hikers or ramblers even if it could be shown that the occupier was aware of their potential presence on the date or dates of crop spraying.

Negligence

4.16. Anyone beyond the boundaries of the occupiers'land, but nonetheless claiming to have been affected by crop spray, could invoke the tort of negligence against the person responsible for the use of the pesticide(s) concerned.

4.17. In order to invoke a 'duty of care', a person must show that he or she was a foreseeable, proximate (that is, closely and directly affected) claimant in respect of whom it would be fair, just and reasonable to extend a duty of care. [6]

4.18. Negligence is not a tort of strict liability. It requires fault, that is, a failure on the part of the defendant to exercise the level of care that could be expected of the 'reasonable man'in the position of the defendant.

4.19. In any case in which the defendant has complied fully with the Control of Pesticides Regulations 1986 – both in terms of being properly licensed and in terms of their actual use – it may be especially difficult to demonstrate breach of the duty of care to the claimant. If the defendant could also show compliance with the informal Green Code, this would again tend to indicate the absence of a breach of duty.

4.20. Conversely, any claimant who can establish a breach of the regulations or Green Code will be very well positioned in relation to his or her attempt to show a breach of the common law duty. Knowledge of the particular substances used – including mixtures of pesticides – and the limits placed upon their use will clearly also be of importance in this context.

4.21. Beyond reliance upon a breach of the regulations, the magnitude of risk (for example, spraying in windy conditions) will be material in demonstrating a breach of duty.

Trespass to Land and Trespass to the Person

4.22. Like nuisance, trespass is a tort to land and does not extend to personal injuries.

Unlike nuisance however, trespass does not extend to a person's enjoyment of land. Also, unlike nuisance, the damage occasioned in a trespass must be 'direct'. Where crop spray falls on, or moves to, a person's land, any damage would not be sufficiently direct to qualify as a trespass. Similarly, to found a claim in trespass to the person, the claimant would have to show that the damage was direct, and also that the defendant intentionally or carelessly caused physical contact via the pesticide in question.

Employers' Liability

4.23. Employers'liability is a sub-branch of the law of negligence in respect of which rather detailed rules have emerged over the years. It is entirely possible that an employee charged with the task of administering pesticides might seek to rely on this avenue of redress if harmed.

4.24. In relation to employers'liability, it is now well settled that an employer is required to provide a safe system of work, adequate plant and equipment (including protective clothing) and adequate supervision. [7]

Breach of Statutory Duty

4.25. Exceptionally, a person suffering damage or injury as a result of a violation of a statute or statutory instrument may be able to sue in tort on the basis of a breach of statutory duty where it is apparent that the obligation or prohibition was imposed for the benefit of a particular class of individuals. [8] This could be argued to apply in the context of residents and bystanders claiming to have been affected by crop spray. On the other hand, the courts have shown a restrictive approach to this exception and it is by no means clear that residents or the various classes of bystander (1. 19) would be able to show themselves to be a specific class for these purposes. For example, under the Control of Substances Hazardous to Health Regulations (COSHH) 2002 it is by no means easy to suggest that the affected persons would fall within a specially protected class of persons: viz agricultural workers or rural dwellers, etc. This is because the regulations apply in many other contexts, not just agriculture.

Congenital Disabilities (Civil Liability) Act 1976

4.26. Under section 1(1) of the Act, a child who is born alive but disabled as a result of an occurrence before its birth may in certain circumstances sue the person responsible.

The Act applies in respect of children born with a 'disability'and liability can only arise when the affected parent would not have been owed a duty of care in tort.

PERVASIVE PROBLEMS IN ESTABLISHING LIABILITY

Causation

4.27. Although tort law is centrally concerned with various duties owed by one citizen to another, and while it is crucial in tort actions to show a breach of these duties, it is also a general requirement that the claimant must show causation:for example, that there was a causal link between the action and the damage suffered by the claimant.

4.28. A major obstacle to the claimant in the context of actions based on the alleged misuse of pesticides is the difficulty of showing causation. As we have discussed in chapter 2, many of the illnesses and medical conditions alleged to have been caused by crop spray are poorly understood. Even where there is a known link between a particular type of illness and the hazard to which the claimant was exposed, requiring the claimant to prove on the balance of probabilities that it was the defendant's wrongful exposure that caused the disease or illness to occur may be more than the claimant can manage.

The Material Contribution to Harm Principle

4.29. Where the claimant can show that the exposure makes a material contribution to harm, he or she will be treated as though causation has been established in the normal way.

4.30. In order to invoke this principle, a resident or bystander would have to be able to identify the precise agent that caused the harm, which may well prove extremely difficult in connection with diseases and illnesses for which the aetiology is only partly established by science.

The Material Contribution to the Risk of Harm Principle

4.31. The full rigour of the material contribution to harm principle may have been softened in recent cases, which have imposed liability when there has been a material contribution to the mere 'risk'of harm. In such cases the 'all or nothing'approach has been abandoned in the interest of justice and the defendant may be liable if there is a known link to only a portion of the damage inflicted upon the claimant.

4.32. Even under the above principles, the courts are likely to be reluctant to grant injunctions to prevent the sprayer making a contribution to harm or the risk of harm in advance of actual damage occurring, except in severe cases. This is because the courts are reluctant to inhibit action where this could have severe economic consequences in the absence of strong proof.

Remoteness of Damage

4.33. A second general principle that runs throughout the law of tort is that the defendant will only be held liable in respect of 'types of harm'that were reasonably foreseeable at the time of the allegedly tortious event.

4.34. Case law in relation to the remoteness principle provides little concrete guidance as to the way in which the courts will decide whether any particular type of harm is foreseeable or not. Given the relative rarity of some of the conditions that have been claimed to be attributable to crop spraying, it is likely that a court would hold attribution to be too remote. The more uncommon the condition, the more likely it is that a court would take this view.

Contributory Negligence

4.35. Contributory negligence is a term applied to a partial defence to tortious liability.

The gist of this defence is that the claimant's loss or injury is partly attributable to his or her own fault and therefore any damages payable in respect of the loss or injury should be reduced to reflect this fact.

Limitation of Actions

4.36. In all tort actions, there are time limits within which a claim must be brought. These are imposed by statute and are known as 'limitations of actions'. In cases involving personal injuries (including illness) and death, the basic limitation period is three years.

This constitutes an additional barrier for those who claim to have been affected by pesticide use more distantly than three years ago.

4.37. The three-year period begins to run, according to section 11(4) of the Limitation Act 1980, 'from (a) the date on which the action accrued; or (b) the date of knowledge (if later) of the person injured'.

RELATED LEGAL ISSUES

4.38. Although not capable of grounding a tort action, there are other legal issues associated with the use of pesticides that may be material in causing a present user to discontinue that use in the future.

Liability in Cases of Contract Farming

4.39. In cases where the farming of a particular site owned by X has been contracted out to Y, a question may well arise as to which of the two parties will be liable under tort law for any damage or loss arising from pesticide use. Notwithstanding the general tort principle of vicarious liability, there is no simple answer to this question. The best way to answer the question is to consider each tort in turn.[2]

Public Health Legislation

4.40. Many of the sorts of conduct that would formerly have given rise to an action in nuisance are now also covered by statutes such as the Clean Air Act 1993 and the Environmental Protection Act 1990. Section 79 of the 1990 Act lists a series of 'statutory nuisances' including emanations of industrial dust or smell, and it is arguable that crop spray would fall within this definition.

4.41. The Health and Safety Executive (HSE) Inspectors deal with grievances in respect of incidents of agricultural spray drift involving residents and bystanders.[3] However, there are a number of questions regarding the effectiveness of the statutory regulations (5. 59-5. 84) and their enforcement, which have meant that the number of prosecutions for spray drift is small (5. 89-5. 94). Some spray operators do not hold licences,[4] but even for those that do, there is no system for the suspension of such licences following a conviction.

Human Rights Legislation

4.42. The Human Rights Act 1998 imposes obligations on public authorities to act in accordance with the rights contained in the European Convention on Human Rights

[2] This has been done in the full report written for the Commission by John Murphy.
[3] The Environment Agency is responsible for pesticide spray drift that pollutes watercourses.
[4] Not all spray operators hold licences. 'Grandfather rights'exempt all those born before 31 December 1964

(Convention Rights). The Act therefore has a 'vertical effect' - it governs the relationships between the state and the citizen.

4.43. Although the courts under the Human Rights Act are themselves defined as public authorities, and are therefore required to develop the law in accordance with Convention Rights, the 'horizontal effect' of the Act is limited. This is because the Act is not primarily intended to govern relationships between private individuals and directly create any new set of rights between them.

4.44. Thus, the fact that crop spray may impact on a claimant's family and private life (respect for which is guaranteed by Article 8 of the Convention) does not necessarily avail that person of a remedy. At the most, in cases such as these to date, the courts have held that any obligation of the state extends only to the provision of information on the pesticides being used.

The EC Environmental Liability Directive (2004/35/EC)

4.45. Although seemingly seeking to expand liability in tort, and specifically covering the subject of pesticides (Annex III tiret 7 (c)), the recent Environmental Liability Directive, despite its title, is not really concerned with legal liabilities between private parties.

The Directive establishes a complex regulatory system to redress 'significant environmental damage', a term given restrictive definitions in the Directive. It requires polluters (operators) to pay prevention and remediation costs, and is not designed as a basis for personal injury recovery. 'Individuals and others who may be directly affected by actual or possible damage, and qualified entities (Non-Governmental Organisations) may request action by a competent authority, and seek judicial review of the authority's action or inaction. '[9]

THE ROLE OF THE LEGAL SYSTEM

4.46. There is little opportunity at present for an individual to seek redress through common law for alleged harm from crop spraying. This is because the law as presently developed requires a clear causal link between pesticide application and harm (or the risk of harm) to the individual. This is often difficult to establish because of the complex nature of health concerns attributed to pesticide exposure (2. 9-2. 12), the limitations of epidemiology (2. 20-2. 34) and toxicology in such cases (2. 40-2. 50), and the inadequacy of the systems that are in place to react to and monitor bystander exposure (2. 70-2. 96). Without such a demonstrable link, and in view too of the limitations of the various possible categories of tort liability described above (some of which are predicated on damage to property rather than to the person), it is unlikely that an individual will be able either to prevent crop spraying by means of an injunction (4. 32), or successfully to impose liability upon the sprayer.

4.47. Given the limitations of the present development of the common law, it seems to us that a more precautionary approach to crop spraying is best achieved through public regulation rather than reliance upon civil suits between individuals seeking monetary compensation for harm already done. Regulation seeks primarily to prevent damage rather than to assign liability after damage has occurred. A degree of regulation is already in place through the establishment of a variety of controls upon spraying, some of which are set out in legislation establishing standards and the machinery for mandatory control of pesticide approval and for the governance of its use. Other nonstatutory regulations are established through voluntary schemes and advice such as the 'Green Code'. The effectiveness of these measures will be evaluated in the next chapter.

REFERENCES

[1] Cornell University Law School. (2005). *Tort law:an overview.* Accessed at:http://straylight. law. cornell. edu/ topics/torts. html
[2] *Sturges v Bridgman* (1879). 11 Ch D 852, at 863.
[3] Occupiers'Liability Act 1957, s 2(2).
[4] *McGeown v Northern Ireland Housing Executive* (1995). 1 AC 233.
[5] Occupiers'Liability Act 1957, s 1(4)(a) (as amended by the Countryside and Rights of Way Act, 2000).
[6] *Caparo Industries v Dickman* (1990). 2 AC 605.
[7] *Wilsons and Clyde Coal Co Ltd v English* (1938). AC 57.
[8] (1982) AC 173, at 186, *per* Lord Diplock.
[9] Department for Environment, Food and Rural Affairs. (2005). *Environmental Liability.* Accessed at: http://www. defra. gov. uk/environment/liability/#1

In: Crop Spraying and the Health of Residents...
Editor: The Royal Commission, pp. 77-110
ISBN: 1-60021-391-X

Chapter 5

THE GOVERNANCE OF AGRICULTURAL PESTICIDES

5.1. In chapter 4 we highlighted the difficulties of seeking redress for exposure to pesticides under common law and emphasised the desirability of a more preventative system of public regulation. This chapter deals with UK and European Union regulation of agricultural pesticides, as well as policy formulation and related initiatives. We examine the pesticides approval process, controls on pesticide use and the enforcement of regulations for agricultural pesticides. In the final section we briefly consider the economic implications of our recommendations.

AGRICULTURAL PESTICIDES REGULATION

5.2. The UK has enacted an extensive range of statutory regulations for the approval, advertisement, sale, storage, supply and use of agricultural pesticides. These have been built up over the years and aim to enable 'effective means of pest and disease control, consistent with protecting the safety of people, animals and the environment'. [1] The principal statutory provisions are listed in box 5A[2] and discussed in more detail later in the chapter.

THE APPROVALS PROCESS FOR AGRICULTURAL PESTICIDES

5.3. At present in the UK there are two systems for the approval of the active ingredients of agricultural pesticides, a European one and a national one. These systems aim to ensure that all pesticides are safe and effective before they are marketed, sold, stored and used. Over time the European system will prevail for the approval of active ingredients. Companies seeking approval for active ingredients are required to submit a scientific data package, the main components of which are set out in appendix K.

The costs of developing and marketing a new active ingredient are typically between £60 million and £100 million. [3]

5.4. Responsibility for the approval and safe use of agricultural pesticides in England and Wales lies jointly with the Department for Environment, Food and Rural Affairs (Defra), the Department for Work and Pensions (DWP) (for the Health and Safety Executive (HSE)), the Food Standards Agency (FSA), the Department of Health (DH) and the National Assembly for Wales Agriculture Department (NAWAD). In Scotland responsibility lies with the Scottish Executive Environment and Rural Affairs Department and in Northern Ireland with the Department of Agriculture and Rural Development, both of whom publish their own regulations. At the European level the responsibility lies with the European Food Safety Authority (EFSA) and the European Commission Health and Consumer Protection Directorate-General (DG Sanco).

BOX 5A.
AGRICULTURAL PESTICIDES REGULATION

The Food and Environment Protection Act (FEPA) 1985.
Section 16 (Part III) of the Act gives Ministers broad powers to:protect the health of human beings, creatures and plants; safeguard the environment; secure safe, efficient and humane methods of controlling pests; and make information about pesticides available to the public.
The Control of Pesticides Regulations (COPR)1986 (as amended).
COPR prohibits the sale, supply, storage, advertisement or use of pesticides unless Ministers approve them and unless general obligations (set out in the Schedules to the Regulations) and conditions specific to individual pesticides (conditions of approval) are met.
The Plant Protection Products Regulations (PPPR) 2003.
PPPR implements EC Directive 91/414/EEC (box 5B) in Great Britain; Northern Ireland has similar legislation. PPPR applies to active substances placed on the European market since July 1993. COPR will continue to apply to all other active substances until they have been reviewed at EC level. PPPR 2003 has been drawn up to: take account of devolution; reflect procedures for representation against regulatory decisions; better accommodate plant protection products containing more than one active substance; and reflect the impact of a decision on whether an active substance can be included in Annex I of EC Directive 91/414/EEC and the action that must be taken.
The Plant Protection Products (Basic Conditions) Regulations 1997.
These Regulations supplement PPPR, applying control and enforcement provisions similar to those of COPR to plant protection products, including general conditions relating to the advertisement, sale, supply and storage, use and aerial application of plant protection products.
The Control of Substances Hazardous to Health (COSHH) Regulations 2002.
These Regulations lay down requirements for the assessment and control of exposure to hazardous substances in the workplace. They cover substances, such as pesticides, that are classed as very toxic, toxic, harmful, irritant or corrosive, those substances with statutory occupational exposure limits, and harmful micro-organisms and dusts.
The Code of Practice for the Use of Pesticides on Farms and Holdings (the Green Code) 1998.
The Code provides advice to assist farmers, growers and other professional users of pesticides to meet their legal obligations and other standards of good practice. The Green Code is authorised by statute but its provisions are not directly legally binding.

The European System

5.5. The approval of agricultural pesticide active ingredients in the UK is increasingly achieved through an evaluation and authorisation system based upon the European regulatory framework for plant protection products (broadly agricultural pesticides) set out in the Plant Protection Products Directive 91/414/EEC. Individual Member States are required to amend their national arrangements and legislation in order to meet the requirements of this Directive. In the UK this has been achieved by the Plant Protection

BOX 5B. KEY FEATURES OF THE EUROPEAN AGRICULTURAL PESTICIDE APPROVALS SYSTEM

Directive 91/414/EEC is intended to harmonise national arrangements for the evaluation and authorisation of plant protection products within the European Union (EU). The Directive requires that all active ingredients used in products for plant protection purposes that were on the market within the EU on or before 25 July 1993 must be subject to a review of their safety regarding human and ecosystem health. New pesticides must also be evaluated and authorised.
Companies seeking to introduce new active ingredients to the EU apply to a Member State of their choice (known as the Rapporteur Member State) for assessment on behalf of the European Commission. The Rapporteur Member State initially ensures that the application dossier is compliant with the requirements of the Directive (a completeness check) and reports its findings to the European Commission. The result of the completeness check is then considered at the Standing Committee on the Food Chain and Animal Health (SCFA) where an opinion on completeness is given. Following detailed evaluation of the dossier, the Rapporteur Member State submits a report of its evaluation (the Draft Assessment Report) to the European Food Safety Authority (EFSA), together with a recommendation as to whether or not the active substance should be approved. [4]
For active ingredients already present in products on the market on or before 25 July 1993[1] the company is required to submit a dossier to support the active substance listed for (re)review to a Rapporteur Member State identified by the European Commission. The Member State initially ensures that the dossier is compliant with the requirements of the Directive (incomplete dossiers are reported to the European Commission). Following detailed evaluation of the dossier, the Rapporteur Member State submits a report of its evaluation (the Draft Assessment Report) to EFSA with recommendations as to whether or not the active substance should be approved.
Once the Rapporteur's Draft Assessment Report has been received by EFSA, arrangements are made for its consideration within the framework of the SCFA, the first stage of which involves discussion within the peer review system managed by EFSA. Following the conclusion of the peer review process a report of the discussions and conclusions of the meetings is prepared for each active substance and circulated to all Member States and other stakeholders for comment and is then discussed and finalised in an evaluation meeting with Member States.
The finalised report from EFSA is then submitted to the European Commission. [5]
Positive decisions of EFSA are published in the Official Journal of the European Communities[6] and approved active ingredients appear in a list known as Annex I. Annex I is being created over a period of time by the review of existing active ingredients and authorisation of new ones. In the light of current scientific and technical knowledge an active substance shall be included in Annex I for an initial period of 10 years if it may be expected that plant protection products containing the active substance do not have any harmful effects on human or animal health, or any unacceptable influence on the environment. [7]
Once an active ingredient has been approved, applications can be made in individual Member States for products that contain it. The approval of such products in the United Kingdom follows a similar process to that for active ingredients.

[1] On or before 1 July 1994 in Austria, Finland, Iceland, Norway, Sweden or Liechtenstein.

Products Regulations(PPPR) 2003 under which, over a period of time, all agricultural and horticultural pesticides will come to be regulated. The key features of the European system are outlined in box 5B.

5.6. Existing active ingredients are currently being reviewed and this process is taking considerably longer than originally expected. The programme is designed to ensure that all available plant protection products are supported by up-to-date information on safety and efficacy. Approximately 850 substances are being handled. Many are not being supported by companies following scientific risk assessments because of the cost of providing additional information; their approvals are therefore being revoked. By the end of 2003 about half the original list of active ingredients had been removed. For the remainder, data still have to be evaluated. Further removals are however likely. [8]

Residents and Bystanders in the European Approvals System

5.7. Under the European approvals process it is assumed that the acceptable operator exposure level (AOEL) can be used as a reference dose against which resident and bystander exposure is assessed (3. 28 and 2. 50). At the European level specific responsibility for questions regarding 'the safety of plant protection products for the user/worker, the consumer of treated products and the environment, as well as plant health'[9] rests with the Panel on Plant Health, Plant Protection Products and their Residues (PPR), a scientific panel of EFSA. Questions are referred to the PPR panel by European Commission-level Working Groups, to whom it responds. The Chair of the PPR panel told us during the course of this study:[10]

> The subject of bystander exposure is not considered by EFSA or my risk assessment PPR panel as requiring priority attention to re-evaluate the risk assessment process. EFSA responds to requests for guidance and risk assessment from the European Commission, the European Parliament and/or Member States. In addition it may also consider self-tasking queries, which arise from its own internal risk-assessment activities, for example, the ongoing review of existing plant protection products. To date we have not received any questions on bystander exposure and we do not see it as a priority. Current risk assessment and risk management procedures at Member State level are considered to be adequate.

5.8. In contrast, the issue of bystander exposure was addressed in 2002[11] by the Scientific Committee on Plants (SCP), the forerunner to the PPR panel. This was in response to a request in 2001 from the European Commission for comment on a Draft Guidance Document *Guidance for the Setting of Acceptable Operator Exposure Levels (AOELs).* [12] The SCP noted that;

> There appears to be no clear definition of bystander. In addition, specific criteria to assess or estimate bystander exposure have not yet been developed. The SCP is of the opinion that a difference should be made between a subject who is at risk of being exposed during the application of the PPP because he is occasionally in the proximity of the field and a subject who lives or works near the field being treated ... For those who will "inevitably" be near the field, the AOEL could, in principle, be applied, although it should be noted that such bystanders include individuals from the general population and not a selected sub-group such as the operators. In conclusion the SCP recommends to the Commission

to reconsider the concept of bystander, their appropriate exposure scenarios and the adequacy of applying the AOEL to this sub-population. [13]

5.9. While this recommendation was formally noted by the Commission and adopted at the Working Group-Legislation meeting of 17/18 October 2002, it does not appear to have been followed up by the European Commission.

5.10. We recommend that the UK government press the European Commission to implement the recommendations on bystanders made by the former Scientific Committee on Plants, taking into account the concerns discussed in our report.

The National System

5.11. Although new pesticide active ingredients are now approved through the European system, existing approvals are maintained through the national systems of Member States, who retain responsibility for approving the final commercial product containing the active ingredients and co-formulants. In the UK, under the Control of Pesticides Regulations (COPR) 1986, the evaluation of scientific data (submitted by companies) on agricultural pesticide active ingredients is carried out by the Pesticides Safety Directorate (PSD).[2]

5.12. The PSD is an executive agency of Defra employing around 200 scientific, policy and support staff at offices in York. The PSD evaluates the evidence provided by the applicant for the approval of agricultural pesticide active ingredients on the basis of safety to people and the environment, and of the efficacy of the product. The PSD is also responsible for advising UK Ministers on all aspects of pesticides approvals policy.

This includes data protection, public access to information, and better regulation. [14]

5.13. The PSD, together with the HSE, provides the secretariats for the Advisory Committee on Pesticides (ACP) and its sub-groups.

5.14. The ACP evaluates new active ingredients and reviews existing ones. The ACP was set up as a statutory committee in 1987 by Ministers under section 16 (7) of the Food and Environmental Protection Act (FEPA) 1985, 'to give them advice, either when requested to do so or otherwise, on any matters relating to the control of pests in furthering the general purposes of Part III of the Act'. [15] These general purposes include 'the continuous development of means to protect the health of humans, creatures and plants, safeguarding the environment, securing safe, efficient and humane methods of controlling pests, and making information available to the public'. [16] Under the Control of Pesticides (Advisory Committee) Order (Northern Ireland) 1987, the ACP was established as the Committee on Pesticides for Northern Ireland.

5.15. The ACP has 19 members with expertise in a range of areas including health, the environment, animal and plant sciences, agriculture and food, along with two lay members. FEPA requires Ministers to consult the ACP regarding:

- regulations which Ministers contemplate making;
- approvals which Ministers contemplate giving, revoking or suspending; and
- conditions to which Ministers contemplate making approvals subject.

[2] Non-agricultural pesticides are evaluated by the HSE.

5.16. The advertisement, sale, supply, storage or use of agricultural pesticides is prohibited unless Ministers have approved them. According to the ACP, Ministers grant approval 'that the product can be used without unacceptable risk to people, livestock or domestic animals and the environment'. [17]As with the European system, once a pesticide has been approved it will be subject to conditions of use and routine review. A pesticide may be reviewed at any time if any evidence emerges concerning its safety. If appropriate, the approval can be restricted or revoked if there are any safety concerns. [18]

5.17. The ACP takes as its starting point the data package evaluations carried out by the PSD (or the HSE for non-agricultural pesticides). The documents put to the ACP by the PSD (and the HSE) identify issues and seek the ACP's guidance. Following discussion by the members with input from scientific experts from departments, the ACP's conclusions and advice are drawn together by the Secretariat and agreed by ACP members. Advice from the ACP goes to Defra, the DH, the FSA, the HSE, the DWP and the devolved administrations in Scotland, Wales and Northern Ireland. Final Ministerial approval for use across the UK cannot be granted without the consent of all these bodies.

Resident and Bystander Exposure in the National Approvals System

5.18. As with the European system the UK government's approach to resident and bystander exposure is based on the assumption that the AOEL can be used as a reference dose against which resident and bystander exposure can be assessed. Specific statements about controlling spray drift are not written on pesticide labels because it is assumed that there is a general legal control over these activities[19] (5. 59 - 5. 88). Resident and bystander exposure was a key topic at an ACP open meeting on 10 July 2002 and has been considered several times at ACP meetings[3] and in advice to Ministers. The PSD consultation on buffer zones and access to information also addressed the issue of resident and bystander exposure.

THE OPERATION OF THE ACP AND THE PSD

5.19. In our Twenty-first Report, *Setting Environmental Standards,* we stated that the handling of risk and uncertainty includes two important components:'looking at broader uncertainties about the current issue which extend beyond the available scientific evidence, and considering a wide range of possibilities; and, where sufficient data are available, quantifying and analysing the risks associated with the issue under consideration'. [20]

5.20. The limitations and uncertainties in any estimate of risk need to be made clear in ways meaningful to people without specialist knowledge. Action taken 'in the name of precaution (from use of worst-case scenarios and safety factors in assessments through to application of the precautionary principle in decision making) should be transparent and subject to review in the light of [the] development of understanding'. [21]

5.21. In 2002 the government issued its own guidelines, *Improving Government's Capability to Handle Risk and Uncertainty*, [22] key messages from which include: [23]

[3] Minutes from ACP meetings can be found at http://www. pesticides. gov. uk/acp. asp?id=743.

- Rising public expectations for government to manage risk are set against a backdrop of declining trust in institutions, declining deference and increased activism around specific risk issues, with messages amplified by the news media.
- Departments and agencies should make earning and maintaining public trust a priority when dealing with risks to the public. There should be more openness and transparency, wider engagement of stakeholders and the public, wider availability of choice and more use of “arm’s-length”bodies such as the Food Standards Agency to provide advice on risk decisions. Underpinning principles for handling and communicating on risk to the public should be published for consultation.

5.22. In addition the government has issued *Principles of Managing Risks to the Public*, [24] which includes five principles for the handling of all types of risk to the public: openness and transparency; involvement; proportionality and consistency; evidence; and responsibility.

5.23. These are the guidelines under which all public bodies, including the ACP and the PSD, are expected to operate.

Advice to Ministers

5.24. After its open meeting in 2002 the ACP concluded that:

> There was a vigorous discussion on the assessment of risk to bystanders at the Advisory Committee on Pesticides third annual open meeting in July 2002. Members concluded that on the basis of the information currently available the risk assessment for bystanders used at present provides adequate protection, even if spray is applied to the edge of a field. The Committee has asked PSD to collect some further experimental data to provide further support to this view. Nonetheless, the Committee recognises that many may consider it socially unacceptable to spray right to the boundary of a neighbour’s property. If Ministers agree, they may wish to consider options to restrict this practice. Similar discussions with respect to people using public rights of way across fields also concluded that current risk assessments provide adequate protection’. [25]

This was also the advice for Ministers given by the ACP to the PSD. [26]

5.25. At a later meeting the ACP considered new data on exposures to spray drift at distances less than 8 metres from the sprayer (*ACP9*(297/2003) – 3. 32-3. 42) and produced further advice in a document entitled ‘Advice to Ministers’which stated that overall, the new information provided in paper *ACP9*(297/2003) did not change the advice on the bystander risk assessment given following the open meeting. This document also reported that the ACP had ‘identified some further confirmatory work to be completed’, which would ‘entail case studies to illustrate average and maximum daily exposures in worst case scenarios as compared with those estimated by the standard approach that is currently adopted in risk assessment’. The ACP concluded that ‘depending on the outcome, a small minority of exposure assessments might need further “refinement”to confirm acceptability’. [27]

5.26. Following a further ACP meeting (301) which discussed paper *ACP20*(301/2003) (3. 32-3. 42), ACP advice in another document labelled ‘Advice to Ministers’stated:[28]

- the current approach used to assess risks is protective, with the possible exceptions of dithianon, trifluralin, and the soil fumigants; however the approach should be expanded to address risks of those exposed over longer periods more transparently, and PSD should take this refinement forward into EU discussions;
- there should not be a movement away from using the AOEL for bystander risk assessment;
- the action that PSD has already initiated to conduct an immediate adverse data review of dithianon was commended;
- the assessment for trifluralin is based on the usually more conservative acceptable daily intake (ADI), and PSD should review the available database to establish if a more appropriate AOEL can be established and whether this indicates an acceptable risk;
- review of the soil fumigants is imminent under the EU review programme and the concerns identified do not suggest a need to act more rapidly than already scheduled for the EU reviews;
- the possibility of funding research to collect air monitoring/biomonitoring data for bystanders should be considered by PSD;
- HSE and PSD should consider the practicality of defining standards or limits for airborne concentrations of volatile pesticides;
- previous advice that buffer zones around residential property are not necessary to protect human health, and the previous suggestion that Ministers may wish to consider options to restrict the practice of spraying right up to the boundary of a neighbour's property for reasons of social acceptability, remain valid; and
- the proposal of five m[etre] set-aside strips in the proposed [Common Agricultural Policy] CAP reforms should be supported, as these could be useful in this context.

5.27. We understand from the PSD that:

> Advice from the ACP comes forward in written form from the Secretariat of the ACP to officials in Departments. The comments from the ACP can include specific advice as well as more general advice such as that given on the possible social unacceptability of spraying.
> Once received, the advice is acted on directly by officials on behalf of Ministers – this is the case for technical issues and routine regulatory decisions. Advice may also be included in submissions to Ministers, where additional background and further explanations are provided and key issues for the Ministers attention are identified.
> The advice following meetings 297 and 301 confirmed the earlier advice given to Ministers that buffer zones around residential properties are not necessary to protect human health, and the previous suggestion that the Ministers may wish to consider options to restrict the practice of spraying right up to the boundary of a neighbours' property for reasons of social acceptability. This advice was not forwarded in a formal submission to Ministers – being confirmed by earlier advice – but would have been mentioned in oral briefings to Ministers. The technical advice provided was actioned by officials. [29]

5.28. We recognise that it is not always practicable for all advice to go to Ministers themselves, particularly when it is highly technical and routine. We do however believe that it is confusing for all advice to be labelled as 'Advice to Ministers' when some of it is intended for officials only. In our opinion it should be the decision of the committee whether its advice is intended for Ministers or for official use. We recommend that advice from committees should be differentiated between advice to the Minister, written in accessible language, and advice to departments, which would normally be for officials to handle.

5.29. When Defra announced that it was asking the Commission to carry out the resident and bystander exposure to pesticides study the Minister said 'The Government's top priority is to ensure that the safety arrangements we have in place protect the public. The independent scientific advice to me is very clear that the existing system provides full reassurance on that score. '[30]

5.30. We asked the PSD exactly what advice was given by them to the Minister regarding risk and public concern, and were told that 'PSD and the ACP has been advising Ministers since this issue [resident and bystander exposure] was first raised at the Open Meeting of the ACP in 2002. If your question concerns whether Alun Michael was made aware of the issue of public concern and perception of risk then we can confirm that Ministers were firstly made aware of the ACP's view that, although they had identified no scientific concerns, those living in rural areas might have a perception that they were at risk. This was the basis on which the consultation was issued. At the end of the consultation period Ministers were made aware, via a submission, of the views expressed in the responses, including reports that individuals were at risk and in some cases claimed that their ill health was due to exposure to pesticides'[31]

5.31. When we asked the Chair of the ACP about statements of the kind that there are 'no scientific concerns', it was made clear that the ACP has not been able to identify a toxicological cause of ill health. [32] Evidence from the DH adviser to the ACP supported the ACP's position that toxicology testing is robust. [33] The DH adviser stated that confidence in the toxicological assessment appears to be based on the assumption that there is a 'correlation between occupation and bystander effects', the 'perception that the proportion and number of individuals who actually experience adverse effects is relatively low', and the 'heavily assessed'nature of pesticides. [34]

5.32. In chapter 2 we noted the different health effects that residents and bystanders have attributed to exposure to agricultural pesticides, and discussed the systems in place in the UK for reporting and monitoring these effects. In chapter 3 we noted that the exposure assessment was based on considering a narrow set of conditions and needed to cover a wider range and to be better based on actual measurements from field trials. We have not concluded that pesticides have been the cause of the ill health reported but we have not been able to exclude this possibility. We do not agree that the evidence can lead to the unequivocal conclusions that the system provides 'adequate protection'(see 5. 24) and that there are 'no scientific concerns'(see 5. 30), or that it provides 'full reassurance'(see 5. 29) to the Minister.

Setting Environmental Standards

5.33. The PSD consultation on buffer zones emphasised that the risk assessment provided 'adequate protection'. [35] When announcing our study, Defra explained:'The aim of the regulatory system is to protect the health of all people who may come into contact with pesticides, whether they receive only a short term exposure, or exposure over a longer term.

This does not mean protecting 100% of people from every conceivable adverse effect, but rather the aim is to protect people from serious effects. '[36]

5.34. The concept of adequate protection has ethical, economic and social dimensions.

Implicit in some recommendations and decisions about the protection of residents and bystanders from pesticide exposure is an assumption that small risks, or risks to a small sub-set of the population, are outweighed by the benefits of pesticide use. But it is not always explicit who benefits and who might be placed at risk, or what criteria might be employed to establish the acceptability of risk or the appropriate balance of risks with benefits. In our view such judgements should always be clear and explicit whenever decisions of this kind are made.

5.35. Further, as we have argued in our Twenty-first Report, *Setting Environmental Standards*, there are alternative frameworks within which the acceptability of risk might be judged.

For example, society is becoming more committed to the concept of individual rights and will no longer lightly condone interference with the rights, interests and legitimate expectations of any individual. Under a rights-based approach, health and life would qualify for special protection. Such a framework might therefore demand a more precautionary approach in the case of potential exposure and harm.

5.36. Government guidelines stipulate that Ministers and senior officials should take a clear lead in improving risk handling. The Government document on handling risk says 'Increasingly they [scientific advisory committees] have to frame their advice to take account of social and ethical issues and public and stakeholder concern. '[37] We were told by the Chairman of the ACP that:

> The Government sets the overall tone of precaution that it would like to operate. We then try to ensure that we work within that framework and achieve a consistent approach to different pesticides ...

and that:

> We do that [make recommendations for risk management] within a framework of value judgements that has been set by Government and which it is open to the Government to change at any stage. [38]

When we asked the ACP for this 'framework of value judgements'we were given a series of examples of Ministerial responses to specific ACP advice, which was in effect seen by theACP to constitute a framework by default.

5.37. We recommend that the advice to Ministers from technical and scientific advisory committees, such as that of the Advisory Committee on Pesticides on the approval of pesticides, should take account of all the criteria employed regarding risk and precaution, the choices available, and the likely impacts on those affected by the decisions being recommended. This information needs to be reflected explicitly when communicating risk management decisions to the public. The meaning of such terms as 'adequate protection' should be clearly defined and communicated. Areas of scientific uncertainty, ignorance and indeterminacy need to be recognised and their nature described explicitly. It should be clear how uncertainty has been managed within the assessment, for example, through the use of probabilistic analysis (see 3. 53), uncertainty factors, etc. The effect of uncertainty on the estimate of risk should be clear.

The Full Range of Opinion in Scientific Advisory Committees

5.38. The government *Code of Practice for Scientific Advisory Committees* stipulates that the chair of an advisory committee has responsibility for ensuring that:[39]

- every member of the panel has the opportunity to be heard and that no view is ignored or overlooked, using, where appropriate, a structured process which ensures that all views are captured and explored;
- the full range of scientific opinion, including unorthodox and contrary scientific views, are appropriately taken into account; and
- any significant diversity of opinion among the members of the panel is fully explored and discussed and if it cannot be reconciled is accurately reflected in the report and in any other communications with sponsoring departments.

The Code also stipulates that members and secretariats should also ensure that the committee has the opportunity to consider contrary scientific views, and where appropriate, the concerns and values of stakeholders before a decision is taken. [40]

5.39. Minutes[41] of ACP meetings reveal that debate concerning resident and bystander exposure to pesticides did take place and that inevitably concerns were voiced about a number of issues including: the availability of information on adverse health effects from pesticides reported, for example to the National Poisons Information Service; exposure to children; and the effects of mixtures. An alternative approach to the assessment of risk for neighbours and the possibility that safety margins did not apply in a minority of worst-case scenarios were discussed and further research was recommended to confirm the adequacy of the current risk assessment. Sympathy was expressed with the view that the current exposure assessment models had some inadequacies, for example the modelling of oral uptake from hand to mouth.

5.40. Correspondence[42] between members of the ACP as to whether these concerns invalidated the concluding advice to Ministers indicates that members did not dissent from the advice at the time, but at least one member subsequently did. Moreover a number of members repeated their concerns and queried their representation (at the meeting and in minutes). Others drew a distinction between science and opinion and stressed the importance of contributions from members with relevant technical expertise.

5.41. On the topic of ACP members commenting on areas outside of their expertise minutes also reveal:

> The Chairman spoke to Members about dealing with the media. He reminded them that while there was no problem in their speaking to the press, they should make it clear that they spoke personally and not on behalf of the Committee. He told Members that he did not expect them to make public comments on the workings of the Committee without first raising the matter with him, and if necessary, the PSD Chief Executive and then the Minister. He then pointed out that as the membership of the Committee was in the public domain, it was important that when making public statements about pesticides, members

did not comment on scientific issues which lay outside their individual area of expertise. If they did this, there was a danger to the scientific credibility of the Committee. [43]

We find this approach surprising, given the uncertainty in the scientific data concerning a possible causal relationship between pesticide spraying, exposure and human health. Such an approach also calls into question the value of deliberation in scientific committees and the role of lay members, if expertise is always to be privileged in this way.

5.42. Genuine and significant differences of opinion between members of an advisory committee should be reflected in its public statements and in advice to Ministers. In our opinion, in this instance, the assurance given to Ministers represented too sanguine a view of the robustness of the scientific evidence.

5.43. We recommend that advice passed to Ministers from expert committees and statements to the public should properly reflect the full range of opinion within such committees.

The PSD: The Handling and Communication of Risk and Uncertainty

5.44. After the ACP open meeting in 2002, its later meetings and its advice to Ministers on resident and bystander exposure, the PSD launched consultations on buffer zones and access to information.

5.45. In the first consultation the question was posed as follows:'The imposition of no-spray buffer zones is not justified solely on scientific grounds. What, if any, public interest justification do you believe there is for introducing them?'[44] The risk assessment and guidelines were presented as a 'robust system', although it was conceded that 'spraying immediately adjacent to residents'boundaries might in any case be considered socially unacceptable'. [45]

5.46. The consultation exercise appears to have been framed in a way that limited the parameters of the debate. It set aside any uncertainty in the risk assessment and in the toxicological testing, thereby effectively positioning the science as beyond doubt.

Responses to the PSD consultation and other evidence submitted in the course of our study show that many of those who believe that their health has been affected by pesticides felt that the consultation was flawed, as it was founded on a premise that they did not share. In their view, the case for buffer zones and other restrictions was based on a genuine concern about their health. Framing the issue so that the science appeared not to be in question has led to controversy and fuelled mistrust of the PSD for some.

5.47. We recommend that in consultations and in dialogue with the public concerning pesticides, as in other areas of uncertainty, science should not be positioned without argument as being beyond doubt. It should be presented in a manner that is open about the level of risk and uncertainty involved.

5.48. The responses to the consultation on access to information could not initially be made freely available to the public, or to us, because the standard paragraph seeking permission to do so had been omitted from the letter inviting evidence. The invitation stated that 'the purpose of this preliminary consultation is to identify key issues that may be used to inform a formal consultation process'. It was not made clear to the public that this would mean they would not be given access to the responses. Nor has a subsequent 'formal consultation process'been undertaken. Conducting a consultation in this manner runs against government guidelines (regarded as binding on UK departments and their agencies), which stipulate that individual responses should be made available to anyone who asks for them. 'Failure to make material available may be incompatible with Open Government and Freedom of Information provisions. '[46] At the very least this approach appears to have once more exacerbated the level of mistrust between some members of the public and the PSD.

5.49. In accordance with government guidelines for best practice we recommend that consultation responses should be made available to the public, unless the respondent has said that they wish their response to remain private.

Delivery and Policy

5.50. The PSD combines both delivery of the pesticide approval process and policy advice to Ministers on pesticides. An executive agency of the government, the PSD is funded by government for its policy work, however the full costs of evaluating applications for pesticides approval are recovered from the industry through fees and levies.

In the year 2003/04 the PSD received £4. 363 million from the levy for regulatory work which includes monitoring and compliance and £2. 791 million in industry fees for evaluating applications. [47]

5.51. The 'PSD is also responsible for advising Ministers on the development and enforcement of pesticide policy and legislation, and advises Ministers on all aspects of pesticides approvals policy'. [48] In the year 2003/04 £5. 379 million was received from Defra for policy-related activities. [49] The Commission has previously concluded that Government bodies should not hold responsibility for policy and for its execution on the same issue, and it is widely held that these functions should be separated between a government department and an arm's length executive agency or non-departmental public body (NDPB). [50] There is a danger of a conflict of interest, which may be greater where funding is derived from outside government.

5.52. More recently Lord Haskins recommended that policy and delivery should be separated: 'Defra's prime responsibility should be the development of policy, and it should arrange for the delivery of its policies through national, regional and local agencies. Policy and delivery functions should be managed separately so that accountability for policy and delivery is clearly defined'[51] and that 'it is hard to establish accountability for failure (or success) when those who develop policy also take responsibility for its delivery'. [52]

5.53. The key delivery activity of the PSD is the appraisal of candidate pesticides however the aims and objectives of the PSD are much broader:[53]

- to ensure the safe use of pesticides for people and the environment;
- to reduce the negative impacts of pesticides, encourage reductions in their use, taking account of best practice, and the development and introduction of alternative ways to control pests, weeds and diseases. This is part of the move towards sustainable food and farming; and

- to harmonise pesticide regulation within Europe and provide a level playing field for crop protection.

On occasion the importance of health and environmental considerations appears to be subordinate to pest control. The draft National Pesticides Strategy, which we discuss below, is a case in point.

The National Pesticides Strategy

5.54. The National Pesticides Strategy has been prompted by proposals for a forthcoming EU Thematic Strategy (due in 2005) which asks Member States to draw up national plans 'to reduce the impacts of pesticides on human health and the environment, and more generally to achieve a more sustainable use of pesticides as well as a significant overall reduction in risks and in the use of pesticides consistent with the necessary crop protection'. [54]

5.55. According to the PSD, the National Pesticides Strategy 'will help encourage farmers to further minimise their use of pesticides by providing best agricultural and environmental advice and guidance' and 'the Strategy must achieve a balance between adequate crop protection on the one hand, and reducing their adverse environmental impact on the other'. [55] Sustainable use of pesticides is defined as 'minimising the hazards and risks to the environment from the use of plant protection products without putting necessary crop protection at risk'. [56] This implies that 'necessary crop protection' is not to be put at risk while hazards and risks to the environment are to be 'minimised'. Public health is hardly mentioned at all in the draft Strategy.

5.56. We recommend that the draft National Pesticides Strategy should give full recognition to the need to safeguard human health and the environment.

5.57. The positioning of both delivery and policy responsibilities within the PSD appears to have made it more difficult to address health issues relating to pesticide exposure. Although health issues are important there are no medically qualified staff at the PSD. The responsibility for developing the best policy for pesticides is one for the government and should involve departments concerned with protecting the environment and human health, as well as those concerned with pest control.

This function needs to be clearly separated from the approval of pesticides.

5.58. We recommend that responsibility for pesticides policy should be separate from that for the approval of pesticides. One way to do this might be to move the policy function from the Pesticides Safety Directorate to a unit within the Environment Directorate of Defra. The unit should have a clear remit for recognising the importance of safeguarding human health and the environment as well as the requirements for pest control; the Department of Health should have an explicit role in agreeing its remit and assessing its performance. The remaining functions relating to the approval of pesticides could be transferred to the Environment Agency to enable better co-ordination with wider environmental objectives.

The Use of Agricultural Pesticides: Statutory Requirements and the Green Code

5.59. The regulations for pesticides state that 'no person shall use a pesticide unless the Ministers have jointly approved it'. [57]Approval is based on the expectation that pesticides are only used in accordance with their conditions of use. [58] Specific conditions appear on the product label and general conditions in the statutory regulations.

5.60. As noted in box 5A the use of agricultural pesticides is governed by legislation under which various regulations have been made, including: the Control of Pesticides Regulations (COPR) 1986 (amended 1997), the Control of Substances Hazardous to Health Regulations (COSHH) 2002, and the Plant Protection Products Regulations (PPPR) 1997 (amended 2005). Requirements of these regulations pertinent to the use of agricultural pesticides include:

- any person who uses a pesticide (prescribed plant protection product) shall take all reasonable precautions to protect the health of human beings, creatures and plants, safeguard the environment and in particular avoid the pollution of water. [59]
- no person who was born later than 31st December 1964 shall use a pesticide approved for agricultural use unless he has obtained a certificate of competence recognised by the Ministers, or he uses the pesticide under the direct and personal supervision of a person who holds a certificate. [60]
- no person shall use in the course of a commercial service a pesticide approved for agricultural use unless he has obtained a certificate of competence recognised by the Ministers, or he uses the pesticide under the direct and personal supervision of a person who holds a certificate; [61]
- it shall be the duty of every employer to ensure that any person in his employment who may be required to use a pesticide during the course of his employment, is provided with such instruction and guidance as is necessary to enable that person to achieve the standard of competence recognised by the Ministers, and otherwise to comply with the requirements in and under these (COPR) Regulations; [62]
- any person who uses a pesticide shall confine the application of that pesticide to the land, crop, structure, material or other area intended to be treated; [63] and
- where a duty is placed by these (COSHH) Regulations on an employer in respect of his employees,[4] he shall, so far as is reasonable practicable, be under a like duty in respect of any other person, whether at work or not, who may be affected by the work carried out by the employer.[5] This includes: An employer shall not carry out work which is liable to expose any employees to any substance hazardous to health unless he has made a suitable and sufficient assessment of the risk created by that work to the health of those employees and of the steps that need to be taken to meet the requirements of COSHH. The risk assessment shall include the level, type and duration of exposure. [64]

[4] The Regulations also apply to a self-employed person except that Regulation 10 (monitoring) and 11 (health surveillance) do not apply.

[5] Except under Regulations 10 (monitoring), 11 (health surveillance), 12 (1) (information), 12 (2) (training) and 13 (dealing with accidents).

5.61. The *Code of Practice for the Safe Use of Pesticides on Farms and Holdings* (the Green Code), [65][6] provides guidance to farmers and growers engaged in commercial crop production in the UK on how to comply with the requirements of the statutory regulations mentioned above. Advice on good practice is also provided. Failing to follow the sections of the Green Code that relate to the statutory regulations does not in itself lead to criminal or civil liability as the Green Code is advisory only and is not directly enforceable in law. However, the Green Code explains those aspects of the law that are enforceable and where a person is prosecuted under FEPA, or other relevant legislation, a failure to observe the provisions of the Green Code may count against them. Membership of voluntary farm assurance schemes (5. 113-5. 115) requires compliance with certain aspects of the Green Code. Agronomists also advise on aspects of the Green Code (5. 116-5. 117).

5.62. The legal force of the Green Code is unclear and variously interpreted. Its design is confusing and the advice sometimes vague. Advice regarding crop spray does not adequately reflect the importance of its control and therefore needs strengthening.

Strengthening the Code and Reducing Bystander Exposure

5.63. It is not always possible to relate incidents of drift and exposure to specific causes.

Crop spray drifting on to gardens, houses and public rights of way, sometimes directly exposing people to pesticides, may be caused by a number of factors including spraying using an inappropriate boom height, nozzle or pressure; incorrect, faulty or poorly maintained equipment; excessive vehicle speed; inappropriate wind speeds and unstable local atmospheric conditions (3. 12-3. 16). It may also be caused by off-target application.

5.64. Factors affecting spray drift and bystander exposure are addressed below under the categories in which they appear in the Green Code, namely:training and certification; planning and preparation; working with pesticides and protecting the public.

Training and Certification

5.65. Currently the law requires that those born after 31 December 1964, who apply agricultural pesticides commercially, must hold a Certificate of Competence. Those born before this date are exempt[7] and are said to hold 'Grandfather rights'. According to the Central Science Laboratory survey of spray practice, 9% of the UK arable area was sprayed under 'Grandfather rights'in 2004, representing 31% of holdings. [66]

5.66. Certificates of Competence are awarded on successful completion of a practical test set by the National Proficiency Tests Council (NPTC) in England and Wales, or the Scottish Skills Testing Service (SSTS). Because the required spraying skills vary widely between the various types of application work, candidates are assessed under the one or more modules that are most appropriate for their professional work. Certificate holders who change their work and use different application equipment may need to obtain a new certificate under a more

---

[6] The Green Code is currently being revised and the new code is expected to come into force in mid-to late 2005.
[7] Unless they are providing a commercial service, in which case they must have a certificate.

relevant additional module. Certificates are not time limited and there is no statutory power enabling Ministers, the courts or the NPTC to revoke or suspend certification.

5.67. To gain a Certificate of Competence, spray operators must be able to demonstrate competence and understanding in a number of areas including: legislation and codes of practice relevant to the use of pesticides; how to read product labels in order to ensure correct use; being able to identify the hazards and risks associated with the use of the product and the necessary precautions to take to minimise them; how to keep proper records and knowledge of the role of health surveillance and health monitoring. [67]

5.68. We support the need for training those involved in the use of pesticides along the lines of the NPTC and SSTS. Lack of knowledge and training in the areas listed above is likely to lead to increased risk of pesticide exposure for humans and the environment. We are concerned that anybody using agricultural pesticides without appropriate training and recognised certification may not be aware of the requirements of safe use or have the necessary skills to use pesticides safely. Anyone using agricultural pesticides must be fully trained. We recommend that 'Grandfather rights', allowing those born before 31 December 1964 to spray without passing the relevant tests, should be phased out by 2010.

5.69. At present farmers, even those not qualified through the training and certification process, can buy agricultural pesticides for use on their own farm. We recommend that the purchase of agricultural pesticides should be restricted to those who are permitted to use them.

Planning and Preparation

5.70. The COSHH Regulations require that exposure of anyone, including members of the public, who may be affected by work involving substances hazardous to health be either prevented or, where this is not reasonably practicable, adequately controlled. This can be achieved by prevention (for example, substitution, alternative pest control), by technical, engineering or operational controls, and by the use of personal protective equipment.

5.71. The Green Code recommends that spraying equipment is cleaned, repaired or adjusted[68] and that control measures (e. g. equipment) are used and maintained in an efficient working order and in good repair. This will usually mean regularly carrying out visual checks, inspections, checking fault detection equipment, and regular preventative servicing. [69] Although the emphasis is primarily on personal protection equipment, other equipment such as nozzles, folding booms and broadcast air-assisted sprayers are also included. Ill-maintained spray equipment can increase spray drift and the risk of bystander exposure. [70]

Working with Pesticides

5.72. It is a statutory obligation to confine the application of pesticides 'to the land, crop, structure, material or other area intended to be treated'. [71]This is generally understood to mean that the act of application must be confined and not the pesticides themselves, as the latter would imply an absolute prohibition of spray drift (which is clearly not what is happening at present and would be very difficult to achieve). It is also a statutory obligation that 'any person who uses a pesticide (prescribed plant protection product) shall take all reasonable precautions to protect the health of human beings . . . [72] Advice on how to comply with these obligations is outlined in the Green Code. General advice is given that 'users of pesticides must ensure that all reasonable precautions are taken to prevent spray drift'. [73] Specific guidelines are provided for weather conditions, and it is recommended that nozzle and pressure choice, vehicle speed, boom height, and equipment maintenance and settings are also taken into account. The general advice to prevent spray drift, the specific guidelines and the recommendations are not statutory requirements in their own right.

5.73. Our discussions with the Silsoe Research Institute and others indicate that factors such as wind speed, boom height and vehicle speed have the potential to influence spray drift significantly. In chapter 3 we highlighted how the modelling of exposure has not been validated by extensive field testing to assess the effect of these influences and to quantify what conditions should be placed on these factors. Nevertheless, and until the further trials we have recommended (3. 53) are completed, the advice in the Green Code represents a pragmatic synthesis of current knowledge on measures to reduce drift and exposure. We believe that adherence to some of the recommended conditions under the Green Code should become statutory duties. These include maximum wind speed, spraying practice as specified on the label, boom height and vehicle speed. The statutory duties should specify the following:

- spraying should not take place in wind speeds greater than a value that should be specified, (for example, 9. 6km/h, the value above which spraying is currently said to be 'inadvisable'). Wind speed and direction at a standard height (2 metres) should be recorded every hour;
- flow rates should not be reduced below the range specified on the label or otherwise a specified rate (for example, 200 litres/ha as assumed in the ACP arable calculations) and should be recorded;
- the boom height above the crop should be carefully regulated to the appropriate level, typically 50cm for most systems;
- under calm conditions it should be noted if there is any sign of near-surface mist, as this is likely to indicate a thermal inversion;
- the speed of the spraying vehicle should not be greater than a value that should be specified (the trials on which the ACP calculations are based were for a tractor speed of 8 km/h). This should apply to all forms of vehicle, including all terrain vehicles (e. g. quad bikes); and
- spraying technology including nozzle type should be as advised on the label or otherwise decided and should be recorded.

5.74. Further research on refining the resident and bystander exposure model should lead to recommendations for revised spraying conditions for all factors relevant to minimising exposure, and thus to a revision of the statutory obligations.

Protecting the Public

5.75. The Green Code states that, 'The COSHH assessment should cover any additional measures required when spraying near premises where there are vulnerable groups of people. Where such people could be affected the assessment should take into consideration any need to notify or otherwise warn anyone of the work activity. '[74] The public need for notification and access to information about pesticide spraying is related to three main areas:

- prior notification, on request, of intended spray activity (what is being sprayed, where and when);
- the right to obtain records of spray activity; and
- notices at the time of spraying on public rights of way and access points for areas, such as in 'right to roam'areas, including all land in Scotland, where the public can walk, to indicate what is being sprayed in the field at that time and any information about whether it is inadvisable to re-enter and if so for how long.

5.76. Residents and bystanders have not always been notified of spray operations and have been exposed to spray drift in different circumstances (for example, in the garden or when walking on public rights of way). Some indicated to us that had they known of spray operations, they would, for example, have shut doors and windows, avoided hanging washing out, not gone for a walk, or even left the area. Some farmers already voluntarily give prior notification to residents and bystanders (by post, fax, phone, email or signs in fields), whilst others do so only in line with legal obligations (for aerial spraying or for specific pesticides, e. g. sulphuric acid (1. 30)).

5.77. The Green Code states that:

> ... the application of pesticides by conventional field crop sprayers should not pose a significant risk to persons outside the boundary of the field being sprayed, provided that the general advice in the Green Code is followed. An assessment may conclude that notifying neighbours may not be necessary for compliance with COSHH. However prior notification of spraying for occupants of land, premises or houses close to the target area is good practice and may help to alleviate concerns they may have on possible ill-health effects. [75]

5.78. Under FEPA, Ministers are given broad powers to make information about pesticides available to the public and, following the consultation on access to information, a pilot study was announced to explore how residents living next to farms could be notified ahead of spraying (1. 6). The study is looking at the practicality and cost of various options for notification and the conclusions are due to be sent to Ministers in late summer 2005.

5.79. We recommend that the residents living next to fields that are to be sprayed be given prior notification of what substances are to be sprayed, where and when. The results of the pilot study in this area announced by the government should be treated as an exercise to determine how best to provide information, not as an opportunity to re-examine the principle of doing so, which should be accepted.

5.80. We recommend that for those wishing to access land, notices should be provided on rights of way or at points where access is likely. These notices should specify what is sprayed and when. A simple sign saying land may be sprayed is not satisfactory, not least because it could be a disincentive to those wishing to access land at times when spraying may not have taken place for some time.

5.81. Following the consultation on access to information, the government also announced 'new legal measures requiring farmers and growers to keep records of pesticides sprayed on crops and to make those records available to the public via a third party' (1. 6).

We endorse these measures and EC Regulation 852/2004 which introduces a statutory obligation to keep records (1. 33). It should not be unduly onerous to keep records of spraying activity. Farmers already do this for the application of all veterinary medicines to livestock.

5.82. Public unease that agricultural pesticides might impact on human health as a result of crop spraying is amplified by the current lack of direct access to information about the chemicals that are contained within the pesticide spray. The PSD consultation on access to information suggested third party notification. However, we have received evidence that this is not a service the General College of Practitioners is keen to see GPs provide. [76] This process could also be costly to the resident or bystander if provided via a solicitor. Our evidence also suggests that third party access would delay information about the pesticides to which the resident or bystander was exposed. This could have significant effects for the treatment of exposure. More than 20 years ago in our Tenth Report, *Tackling Pollution – Experience and Prospects,* we expressed the view that, 'To deny access to data on the grounds that the "public"is not competent to make "correct" use of them is neither a tenable nor an acceptable position'. [77] We remain of that view.

5.83. The Green Code also advises that 'if someone suggests that they have been affected by a pesticide, it is most important to give them or their advisers full and accurate information as soon as possible. That means the full name of the pesticide (including prefixes and suffixes) and any other information which may be needed for treating people or animals, or for a decision to be taken on whether crops will be safe to eat'. [78]

5.84. We recommend that records of which pesticides, and when and where they have been used, should be directly available from the persons responsible for crop spraying upon request to any resident or bystander and to researchers investigating the health effects of resident and bystander exposure.

Buffer Zones

5.85. The Green Code advises that particular care is needed when applying pesticides where there are vulnerable groups of people (for example, near hospitals, schools and retirement homes). Measures to prevent or control any risks to vulnerable people identified by the COSHH assessment may include leaving an untreated area next to the neighbouring property. [79] Currently there is no statutory requirement for an unsprayed buffer zone between a treated area and neighbouring property, even though spraying up to the boundary will increase the likelihood of spray drifting onto that property.

We highlighted the proximity of some residents to sprayed fields in chapter 1 (figures 1-II and 1-III) and noted in chapter 3 that a buffer zone could decrease the extent of exposure from spray drift significantly (3. 24), particularly with active management, such as the use of one metre high vegetation. Until the further research that we recommended in chapter 3 is undertaken (3. 53), setting the width of protective buffer zones will inevitably have to be a pragmatic exercise. In our view, five metres would be a reasonable zone width in line with the approach taken for watercourses. It would also reflect the need to protect children since drift at 0. 9 metres above ground level is initially much greater than at a height of 1. 9 metres, but declines to only slightly higher values five metres away from the sprayer (figure 3-VII).

5.86. We recommend the introduction of buffer zones alongside residential property and other buildings such as schools, hospitals and retirement homes where people may be adversely affected by crop spray. Buffer zones should be evaluated on the basis of the revised residents and bystander exposure model.

Until then, on the basis of our current understanding, we see five metre buffer zones as providing the necessary further protection and encourage the active management of these so as to decrease the likelihood of spray drift affecting residents and bystanders.

5.87. Our recommendation about buffer zones has been formulated primarily in relation to application by boom spray. We have not investigated the issue in relation to orchard crops but would expect the government to develop equivalent protection for those whose property adjoins orchards.

5.88. The likely economic impact of introducing buffer zones is examined later in this chapter (5. 119-5. 126).

MONITORING AND ENFORCEMENT OF AGRICULTURAL PESTICIDES REGULATIONS

5.89. It is impossible to state with certainty the extent to which farmers and growers comply with the statutory regulations and supporting advice in the Green Code.[8] On the evidence we have received, it is clear that many farmers and growers are spraying responsibly, have good relations with residents and bystanders and have never received complaints. Many voluntarily take measures to reduce the risks to residents and bystanders from crop spraying (for example, through prior notification of spraying or the use of buffer zones). However, we received evidence that crop spray on occasion drifts onto gardens, houses and public rights of way, sometimes affecting people directly.

Most residents and bystanders are not notified of spray activity and do not benefit from buffer zones at the edge of sprayed fields adjoining their property. What is clear to us from evidence gathered during this study is:

- not all farmers who spray crops are aware of the Green Code, even though it is sent to all farmers with a holding of more than half a hectare; and
- monitoring and enforcement of compliance with the legal obligations of the Green Code is limited.

[8] However the *Pesticide Survey Report:A Survey of Current Farm Sprayer Practices in the United Kingdom 2004* (Central Science Laboratory 2005) provides a comprehensive insight into crop spraying practices.

5.90. The HSE is responsible for enforcing the law on commercial agricultural pesticides.[9] The HSE Inspectors monitor farms to look for compliance, awareness and good practice according to two benchmarks: pesticide approval conditions and the Green Code.

The HSE receives in total approximately 25, 000 complaints per year (based on 2003/04 figures), approximately 1% of which relate to pesticides. [80] The majority of pesticide-related complaints investigated are from members of the public and relate to spray drift. The HSE reports that these complaints usually arise when crops or land are sprayed when the wind is in the wrong direction and/or wind speed is too high. Notification of intention to spray also remains an issue, particularly in relation to the spraying of sulphuric acid on potato haulms. '[81]

5.91. There are approximately 220, 000 farms registered in the UK. Approximately 10, 000 proactive inspections are made to farms per annum by 70 HSE Inspectors; a reduction from 120 Inspectors ten years ago. [82] Evidence submitted to the Commission highlights the inadequacy of the inspection system, particularly the delay between the reporting of an incident and the Inspector's visit. We have made recommendations for record keeping of pesticide use (5. 84), training (5. 99) and certification (5. 68) and tests for spray equipment (5. 98). We recommend that as part of their on-site visits the Health and Safety Executive Inspectors should request evidence of: record keeping of pesticide use; training and certification; and testing of spray equipment.

5.92. Eight charges for the misuse of pesticides were laid before the courts by the HSE during the year 2003/04; all led to convictions. The average fine imposed by the courts was £1, 824. [83] The number of enforcement notices issued by the HSE Inspectors in recent years under FEPA and COPR is shown in table 5. 1. [84]

Table 5. 1. Number of enforcement notices issued by the HSE Inspectors[85]

YEAR	NUMBER OF ENFORCEMENT NOTICES
2003/2004	66
2002/2003	81
2001/2002	47
2000/2001	98
1999/2000	154

5. 93 Enforcement notices either:

- allow an activity involving pesticides to continue whilst remedial action or preventative measures are carried out within a specified period; or
- prohibit an activity involving pesticides until specified matters have been resolved.

5.94. Currently, there is no specific provision within the legislation enabling Ministers, the courts or the NPTC to revoke or suspend certification of competence (5. 66).

We recommend that courts should be empowered to suspend or revoke the right to spray where operators have been convicted of breaking the law.

[9] The HSE is also responsible for forestry, manufacture and health care. Local Authorities enforce the law in other areas including garden centres, catering and offices.

Other Policies and Non-Governmental Schemes Affecting the Use of Pesticides

5.95. In addition to statutory regulations, the National Pesticides Strategy and the Green Code, a number of environmental policies[10] and non-governmental schemes influence agricultural pesticide use. The key consideration here is what impact they have on resident and bystander exposure.

The Voluntary Initiative

5.96. The government has considered the option of a pesticide tax to reduce the environmental damage from pesticides. This was in part an attempt to ensure that the negative external costs of using agricultural pesticides were reflected in the costs borne by those using them. The idea of a tax was controversial and further consideration has been postponed while the Crop Protection Association (CPA) takes forward a five-year industry-led voluntary programme known as the Voluntary Initiative.

The Voluntary Initiative started in 2001 and incorporates several projects intended to improve practices through the use of technology, information provision and training.

5.97. Under the Voluntary Initiative a commitment was made by the agricultural, horticultural and amenity industries to extend sprayer testing to cover the majority of field sprayers operating in the UK. As a consequence the Agricultural Engineers Association (AEA) and the farming unions, together with other key stakeholders, developed and introduced the National Sprayer Testing Scheme (NSTS).

5.98. Experience has shown that more than half the spraying equipment tested has needed repairs or adjustment to achieve the required standard. [86] By 31 March 2005, 4, 404 sprayers had been tested representing equipment used on 50% of the arable area. [87] Ill-maintained equipment can lead to problems with pesticide application and can increase drift and potentially resident and bystander exposure. We recommend that the National Sprayer Testing Scheme becomes compulsory and that spray equipment should be checked annually so that it is regularly maintained in good working order by both contractors and farmers. Consideration should also be given to applying this recommendation to more rudimentary or small-scale types of spraying, such as those operating from all terrain vehicles (e. g. quad bikes).

5.99. Another important aspect of the Voluntary Initiative is the National Register of Spray Operators (NRoSO). Arable farmers have been urged to sign up to this new professional register that will give official recognition to their. Operators must be able to show that their skills are up to date and, if applicable, must have a Certificate of Competence (NPTC or SSTS). To remain on the register, they must earn enough continuing professional development (CPD) points; typically by attending at least one day's training per year.

---

[10] In addition to the policies we examine, these include:EC Directives *(Water Framework Directive, Habitats Directive, Birds Directive);Waste Management Regulations;*Food Standards Agency Action Plan to Minimise Pesticide Residues in Food; the Strategy for Sustainable Food and Farming *English Nature Wildlife Enhancement Agreement.*

Amongst other responsibilities NRoSO members must give priority to the health and safety of bystanders and consumers. [88] Targets for training were exceeded with 19, 600 operators trained by 1 March 2005, although spray operators accounting for 25% of the arable area have still not joined the scheme. [89] We recommend that membership of the National Register of Spray Operators becomes compulsory and that it should be linked to possessing a Certificate of Competence, which should expire if continuing professional development is not undertaken.

5.100. Recently the House of Commons Environment Food and Rural Affairs Select Committee reviewed the success of the Voluntary Initiative and recommended making the operator registration scheme and the national sprayer testing scheme mandatory. [90]

Set-Aside

5.101. Under the 2003 reform of the EU Common Agricultural Policy (CAP) most farmers seeking to receive the Single Farm Payment (SFP) must 'set aside'some portion of their land. This obligation applies to most farmland other than organic land and permanent pasture. The minimum width of land to set aside has been reduced from 10 metres to 5 metres and the minimum size of a single plot from 0. 1 ha to 0. 05 ha. The current requirements for set-aside land are that it must remain in good agricultural and environmental condition. [91] Orchards are not eligible for the SFP and so are not covered by the set-aside obligation.

5.102. The rules for set-aside indicate that while use of herbicides (and possibly other pesticides subject to being granted a permit) is allowed, it is not a requirement that they must be used to keep the land in good agricultural and environmental condition. Therefore pesticide/herbicide-free buffer zones are possible if good agricultural and environmental conditions can be maintained in other ways. It should be noted that use of spot herbicides to control certain weeds would not give rise to the same exposure scenario as boom or air-assisted sprayers.

5.103. Under the rules governing cross-compliance for the Single Farm Payment scheme, Defra states that:

> Farmers have the option to put all or part of their set-aside land into 6-10 metre set-aside strips, subject to certain restrictions, and only next to controlled water, wet ditches, Sites of Special Scientific Interest, woodland and hedges.

5.104. *Prima facie*, this suggests that the introduction of pesticide-free buffer zones would not qualify for set-aside payments since the locational criteria are primarily defined in terms of the environment rather than human health. However, it should be noted that paragraph 118 of the Defra set-aside guidance note[92] states that it is possible to apply for an exemption from the set-aside management rules for 'reasons of human, plant, or animal health and safety'.[11]

5.105. Defra noted that while the exemption referred to here applies to the rules for managing set-aside rather than the location of the buffer zone itself, this would probably not

[11] See also paragraph 19. 1 of the Scottish Executive Single Farm Payment Information Leaflet 6.

be an impediment to the acceptance of such an exemption.[12] As part of our study Agra CEAS consulted the European Commission on our behalf to ascertain whether this interpretation was correct. The European Commission indicated that if the issue were related to human health it would be within the competence of the Member State to decide on the matter. [93]

5.106. Our recommendation (see 5. 86) that five metre unsprayed buffer zones should become compulsory between areas treated with agricultural pesticides and residential properties is therefore compatible with current opportunities for set-aside land.

Agri-Environment Schemes

Environmental Stewardship

5.107. Environmental Stewardship (ES) was launched in England in March 2005 to replace existing agri-environment schemes.[13] Similar environmental stewardship schemes were launched in Scotland, Wales and Northern Ireland. Environmental Stewardship provides funding to farmers and other land managers in England who deliver effective environmental management on their land. The scheme has three parts. Entry Level Stewardship (ELS) is open to all those farming in England. It is intended to reward good land management and address environmental issues affecting the wider countryside including diffuse pollution, soil erosion and the conservation of farmland birds. Organic Entry Level Stewardship (OELS) is available to those wishing to farm organically. Higher Level Stewardship (HLS) pays more in exchange for more significant environmental benefits in high priority situations and areas. [94]

5.108. In addition to management option relating to large areas of land, such as arable fields, and boundary features, such as hedgerows, management options available under ELS and OELS (but not HLS) include buffer strips of 2, 4 or 6 metres on cultivated land/rotational land or intensive grassland/organic grassland. Defra states that 'there is no reason why every farmer in England should not be part of this scheme in due course'. [95]

5.109. These new agri-environment schemes reflect the current trend in agricultural policy away from subsidising production towards supporting environmental stewardship. This has the potential to reduce crop spraying. In principle ES permits the use of herbicides to control weeds such as nettles, creeping thistle and ragwort, but only by using a weed wiper or by spot treatment using a knapsack sprayer. In some cases farmers and growers will be required to seek prior agreement before they carry out any pest control.[14]

Organic Farming

5.110. Some of those submitting evidence see organic farming on a large scale as a solution to the risks posed to human health and the environment from the use of agricultural pesticides.

---

[12] It was noted, however, that a definitive ruling on this would be needed from the Defra Legal Department. Personal communication between Mr. Brian Bibby, Defra and Agra CEAS.

[13] For example, the Environmentally Sensitive Areas (ESA), Countryside Stewardship Schemes (CSS) and Organic Farming Schemes (OFS).

[14] Derogations are allowed in practice in exceptional circumstances. Where this is the case and farmers or growers need to spray a buffer zone, notification should be given to bystanders along the lines recommended elsewhere in this chapter (5. 80-5. 81).

Organic farming is a system of food production that restricts the use of synthetic agricultural pesticides and fertilisers in favour of a small number of biological pest controls and techniques such as natural nitrogen fixation, encouraging biologically-active soil life, crop rotation, the use of recycled farm manures and crop residues, and appropriate cultivation.[15] The government is promoting the adoption of organic farming for its environmental benefits.

5.111. We commissioned work on the extent to which it might be realistic to expect the majority or indeed all land in the UK to be farmed organically. The report (available on our website) concluded that on the basis of current patterns of consumption, trade and prices there would be a number of obstacles, primarily higher costs and lower yields, which make conversion to organic farming difficult to envisage on a national scale.

5.112. Although organic farming uses significantly fewer pesticides than conventional farming, some pesticides used in organic farming (e. g. rotenone and copper sulphate) have been the subject of human health and environmental concerns (appendix E). The use of pesticides in organic farming is subject to the same regulations as in nonorganic farming and our recommendations therefore apply equally to them.

Farm Assurance Schemes

5.113. These voluntary schemes, which 'have the potential to contribute to a reduction in the environmental impacts of pesticides', [96] evolved in the UK in the 1990s as a partnership between producers and retailers in response to consumer concerns about food safety, environmental protection and animal welfare. In 2004 90% of the UK arable area was covered by one or more crop assurance scheme. [97] Farm assurance schemes produce a series of crop-specific protocols, designed to assure consumers that farmers and growers are producing food in a way that meets all legal standards and to an agreed set of standards of good agricultural practice, which for example can include requirements such as signing up to NRoSO. The schemes include regular, independent checks on member farmers to ensure that they follow the rules. Premium schemes, which set higher levels of assurance, also exist. These include Linking Environment and Farming (LEAF) and organic standards such as those of the Soil Association. Given the purchasing power of major retailers, who often require membership of a scheme for their suppliers, these schemes can exert a powerful and widespread influence on practice.

5.114. Linking Environment and Farming (LEAF) is a Europe wide organisation 'committed to a viable agriculture, which is environmentally and socially acceptable and which ensures the continuity of supply of wholesome, affordable food, while conserving and enhancing the fabric and wildlife of the countryside for future generations'. [98] LEAF encourages farmers to adopt Integrated Farm Management (IFM),[16] to promote the benefits of

---

[15] EC Regulation 2092/91 defines the legal minimum standards of production and processing and the requirements for control and policing of organic farming. The United Kingdom Register of Organic Food Standards (UKROFS) is the government body responsible for implementing this Regulation in the UK. UKROFS license several independent bodies to inspect and certify organic operations and their products. See the EU website http://europa. eu. int/comm/agriculture/qual/organic/reg/index_en. htm.

[16] Integrated Farm Management uses traditional farming methods, such as crop rotation, and modern methods, combined with use of pesticides and fertilisers only when absolutely necessary to keep animals and crops healthy.

this to consumers and to raise awareness of the way many farmers are responding to current concerns, such as those over the use of pesticides.

5.115. The LEAF audit, conducted by LEAF members, checks for a number of practices including the regular maintenance of sprayers within nationally recognised schemes such as NSTS, the use of margins, buffer zones and conservation headlands, and sensitive management of public rights of way. [99] A ten-year review of the audit highlighted good practice among LEAF members, including an increase in the use of uncropped strips to protect hedgerows and watercourses from 40% in 1997 to 65% in 2002. [100]

Agronomist Advice and the BASIS (Registration) Ltd

5.116. Ninety-nine per cent of UK arable area is farmed under advice from agronomists holding the BASIS Certificate of Competence in Crop Protection. Agronomists inspect crops regularly and advise on pesticide choice and, in order to minimise the likelihood of any drift, give guidance on the timing of pesticide application, appropriate weather conditions, as well as nozzle selection, water volumes, LERAP requirements, and set-aside and buffer zone location. Agronomists also encourage spray operators to join NRoSO and have their sprayers tested under NSTS.

5.117. The existence of non-governmental scheme, such as those highlighted above, encourages us in our conclusion that the recommendations in our study are feasible. However the existing voluntary schemes have not proved adequate in themselves to ensure the protection of all residents and bystanders from exposure to pesticides.

The Economic Impact of Our Recommendations

5.118. In the course of this study we have not been able to undertake detailed economic appraisal of all of our recommendations but did commission work on buffer zones and the possibility of a levy on pesticides. We have also noted the recent Beratungsgesellschaft für integrierte Problemlösungen (BiPRO) report for the European Commission, *Assessing economic impacts of the specific measures to be part of the Thematic Strategy on the Sustainable Use of Pesticides*. [101] This provides an analysis of the costs for training, certification and testing of sprayers.

Buffer Zones

5.119. The 2003 Defra consultation revealed concern on the part of farmers about the impact of buffer zones. We asked consultants Agra CEAS to examine the potential economic impact of introducing pesticide-free buffer zones for those parts of farms in the UK where there is potential resident and bystander exposure.[17] The full report is available on the Commission website.

[17] Based on the calculation (1. 22) that the boundary between arable and horticultural and residential property is 20, 700 km

5.120. The effects of introducing pesticide-free buffer zones depend on whether buffer zones are to be cropped or left uncropped. At farm level where the introduction of buffer zones results in relatively small land losses, this will tend to result in a reduction in the enterprise gross margins but will generally leave the fixed cost structure of the farm unchanged. In contrast, relatively large land losses could require a change in farming systems, thereby having a greater impact on the fixed cost structure of the farm. However, for most farms, it is likely that the introduction of pesticide-free buffer zones will require only a relatively small proportion of land to be taken out of pesticide-based production.

5.121. The elimination of pesticide spraying in buffer zones will cut total expenditure on pesticides. It may also result in changes to other variable costs such as fertiliser and seed requirements (particularly where a buffer zone is left uncropped). Production practices may change to offset expected losses from weeds, pests and diseases by, for example, delaying the drilling dates of crops and using a different seed and/or fertiliser rate.

5.122. If a farmer chose to continue cropping in the buffer zone then, as would be expected, the main factor governing the extent of the reduction in yield is the response of the individual crops to the elimination of pesticides.

5.123. At a national level under a 'low input without pesticide scenario' (the highest cost scenario), the cost to British agriculture and horticulture of the introduction of pesticide-free five metre buffer zones on cropped field margins where there is potential for resident and bystander exposure, was estimated to be around £1. 64 million. If 8 metre buffer zones were to be introduced, the cost would rise to around £2. 6 million.

5.124. It is possible that farmers could re-allocate their set-aside allowance to strips adjacent to areas where there is potential for resident and bystander exposure (5. 101-5. 106), in which case the net cost for producers of such crops could be reduced to zero. This would not be the case for producers of top fruit, since they are not under an obligation to set aside land.

5.125. The use of buffer zones for production of crops that would be marketed as organic (on farms which are not organic) would not be possible since the EC Regulations governing organic production require that there be a buffer zone between organic production and conventionally grown crops.

5.126. We recommend that the UK interpretation of the rules on the Single Farm Payment scheme should be changed to allow land in buffer zones to count towards set-aside and that farmers should be strongly encouraged to make use of agri-environment schemes to minimise the economic costs of buffer zones.

Training and Certification of Users

5.127. The BiPRO report highlights the need for the efficient application of pesticides to avoid overuse and associated costs. BiPRO has recommended a mandatory training scheme on a European scale leaving organisational aspects to individual Member States. It was expected that this option would cost European Union farmers 250 million Euros, but would lead to more efficient pesticide use and cut pesticide use by about 9, 000 tonnes. Statutory training and certification would also generate additional income and employment for responsible organisations.

Sprayer Testing

5.128. Ill-maintained sprayers can cause unintended losses and overuse of pesticides. The BiPRo report estimates that making sprayer testing an obligation will involve additional costs for European farmers of approximately 130 million Euros per year but will in the long term realise savings of approximately 9, 000–18, 000 tonnes per year of active ingredients input, equivalent on average to 230 to 460 million Euros. Economic benefits will arise for those running the sprayer tests, and economic costs resulting from adverse environmental and health effects will decline.

Health Costs

5.129. We were told by the Health Protection Agency that our recommendations for improved monitoring of the health effects of pesticide exposure would cost £5-10 million.

This could be met by re-ordering priorities in the health system, but we recognise that this would be problematic. Another option would be to reconsider a pesticides levy. If a levy were to be charged on the value of pesticide sales to farmers then £3. 6 million would be generated for every 1% levied. Extending this to cover the use of pesticides in other areas as well as agriculture (2. 107) would increase this to £5 million.

Conclusions

5.130. We have found a number of inadequacies in the procedures relating to the regulation of the approval and use of agricultural pesticides. 'Advice to Ministers' and public statements made by advisory committees should more transparently reflect the full range of opinion, including uncertainty and value judgements. The public should be more involved and there should be clearer leadership on setting standards of risk and uncertainty. We have recommended separating pesticide policy formulation from the approval of pesticides, a number of additional safeguards in the approvals process and a more cautious approach to resident and bystander risk. We also believe that the time has come to make some of the voluntary standards compulsory.

5.131. We have recommended improvements in the provision of information to residents and bystanders and in measures to reduce potential exposure so as to reflect uncertainty in the science underpinning the risk assessment.

References

[1] Department for Environment, Food and Rural Affairs (Defra) and the Health and Safety Executive (HSE), *A Guide to Pesticide Regulation in the UK and the role of the Advisory Committee on Pesticides* (ACP 19 (311/2005)), www. pesticides. gov. uk/acp_home. asp accessed July 2005.

[2] The source for Box 5A is:Defra and the HSE, *A Guide to Pesticide Regulation in the UK and the role of the Advisory Committee on Pesticides* (ACP 19 (311/2005)), www. pesticides. gov. uk/acp_home. asp, accessed July 2005.
[3] Pesticides Forum, *Paper PF 141:National Pesticides Strategy*, http://www. pesticides. gov. uk/pesticides_forum. asp?id=617#Existing_Pesticide_Regimes_in_Great_Britain_an d_Northern_Ireland, accessed May 2005.
[4] The Pesticides Safety Directorate (PSD), *European Decision Making for New Active Substances and Reviews - An Overview*, http://www. pesticides. gov. uk/approvals. asp?id=247, accessed May 2005.
[5] Ibid.
[6] Official Journal of the European Union, Eur-Lex, http://europa. eu. int/eur-lex/lex/JOIndex. do?ihmlang=en, accessed May 2005.
[7] European Commission, Plant Protection Products Directive 91/414/EEC (Article 5).
[8] The British Crop Production Council (BCPC), *The UK Pesticide Guide 2005*, CABI Publishing, Wallingford, UK, p. 536.
[9] European Food Safety Authority (EFSA), Panel on Plant health, Plant protection products and their residues (PPR), http://www. efsa. eu. int/science/ppr/catindex_en. html, accessed May 2005.
[10] Evidence from Professor Tony Hardy, Chair of EFSA's PPR Panel, 2005.
[11] Scientific Committee on Plants (SCP) (2002), *Opinion of the Scientific Committee On Plants on Commission Draft Guidance Document on the Setting of Acceptable Operator Exposure Levels (AOEL)* (7531/VI/95 rev. 6).
[12] Commission of the European Communities, DG SANCO (2001), *Draft Guidance Document:Guidance for the Setting of Acceptable Operator Exposure Levels (AOELs)* (7531/VI/95 rev. 6), http://europa. eu. int/comm/food/plant/protection/resources/7531_vi_95. pdf, accessed May 2005.
[13] SCP (2002), *Opinion of the Scientific Committee On Plants on Commission Draft Guidance Document on the Setting of Acceptable Operator Exposure Levels (AOEL)* (7531/VI/95 rev. 6), http://europa. eu. int/comm/food/fs/sc/scp/out136_ppp_en. pdf, accessed May 2005.
[14] Defra and the HSE, *A Guide to Pesticide Regulation in the UK and the Role of the Advisory Committee on Pesticides*, (ACP 19 (311/2005)), www. pesticides. go. uk/acp_home. asp, accessed July 2005.
[15] *Food and Environment Protection Act (FEPA) 1985*, TSO.
[16] Ibid.
[17] Defra and the HSE, *A Guide to Pesticide Regulation in the UK and the Role of the Advisory Committee on Pesticides*, (ACP 19 (311/2005), www. pesticides. go. uk/acp_home. asp, accessed July 2005.
[18] Ibid.
[19] Evidence from the PSD, April 2005.
[20] Twenty-first Report, p133.
[21] Ibid.
[22] Strategy Unit, HM Treasury, *Risk:Improving Government's capability to handle risk and uncertainty*, http://www. number10. gov. uk/su/risk/risk/home. html, accessed May 2005.

[23] Strategy Unit, HM Treasury, *Risk:Improving Government's capability to handle risk and uncertainty: Summary of Recommendations*, http://www. number10. gov. uk/su/risk/risk/report/report/summary/ summary. htm#08, accessed May 2005.
[24] HM Treasury, Cabinet Office, *Principles of Managing Risks to the Public,* http://www. hm-treasury. gov. uk/media/C87/A1/risk_principles_180903. pdf, accessed May 2005.
[25] Advisory Committee on Pesticides (ACP), *Assessment of Risk to Bystanders*, http://www. pesticides. gov. uk/acp. asp?id=586, accessed July 2005.
[26] Evidence from the ACP, May 2005.
[27] Ibid.
[28] Ibid.
[29] Evidence from the PSD, April 2005.
[30] Defra, *News Release:Royal Commission to examine science on human exposure risks from crop spray*, http://www. defra. gov. uk/news/2004/040616d. htm, accessed May 2005.
[31] Evidence from the PSD, January 2005.
[32] Evidence from Professor David Coggon, Chair of the ACP, November 2004.
[33] Evidence from John Battershill, Department of Health, February 2005.
[34] Ibid.
[35] The PSD (2003). *A formal consultation document on the introduction of buffer zones between pesticide spray applications and residential properties*, http://www. pesticides. gov. uk/publications. asp?id=431, accessed July 2005.
[36] Defra, *News Release:Royal Commission to examine science on human exposure risks from crop spraying*, http://www. defra. gov. uk/news/2004/040616d. htm accessed May 2005.
[37] Department for Trade and Industry (DTI), Office of Science and Technology (2001). *Code of Practice for Scientific Advisory Committees*, http://www. ost. gov. uk/policy/advice/copsac/cop_01. htm, accessed May 2005.
[38] Evidence from Professor David Coggon, Chair of the ACP, November 2004.
[39] DTI, Office of Science and Technology (2001), *Code of Practice for Scientific Advisory Committees*, http://www. ost. gov. uk/ policy/ advice/ copsac/cop_01. htm, accessed May 2005.
[40] Ibid.
[41] Minutes of ACP meetings 293 (July 2002), 297 (January 2003), 298 (February 2003), 299 (April 2003) and 301 (July 2003), http://www. pesticides. gov. uk/acp. asp?id=285, accessed July 2005.
[42] Evidence from the ACP, April 2005.
[43] ACP, *Final Minutes of the 307th meeting of the Advisory Committee on Pesticides held on 20th May 2004*, http://www. pesticides. gov. uk/acp. asp?id=1204, accessed July 2005.
[44] The PSD (2003). *A formal consultation document on the introduction of buffer zones between pesticide spray applications and residential properties*, http://www. pesticides. gov. uk/publications. asp?id=431, accessed July 2005.
[45] Ibid.
[46] Cabinet Office (2001). *Code of Practice on Written Consultation*, http://archive. Cabinetoffice. gov. uk/ servicefirst/ 2000 / consult /code / consultationcode. htm, accessed April 2005.

[47] Evidence from the PSD, May 2005.

[48] Defra and the HSE, *A Guide to Pesticide Regulation in the UK and the Role of the Advisory Committee on Pesticides*, (ACP 19 (311/2005)), www. pesticides. go. uk/acp_home. asp, accessed July 2005.

[49] Evidence from the PSD, June 2005.

[50] Sixth Report, 428pp.

[51] Lord Haskin's Report, *List of Main Recommendations*, http://www. defra. gov. uk/rural/ruraldelivery/report/text/annex1. htm, accessed April 2005.

[52] Lord Haskin's Report, *Chapter 1 - Institutional and cultural challenges*, http://www. defra. gov. uk/rural/ruraldelivery/report/text/chapter1. htm, accessed April 2005.

[53] The PSD, *About The Pesticides Safety Directorate*, http://www. pesticides. gov. uk/corporate. asp?id=211, accessed April 2005.

[54] Commission of the European Communities, *Towards a thematic strategy on the sustainable use of pesticides* (COM (2002) 349), http://europa. eu. int/eur-lex/en/com/pdf/2002/com2002_0349en01. pdf, accessed July 2005.

[55] The PSD, *National Pesticides Strategy*, http://www. pesticides. gov. uk/environment. asp?id=70, accessed July 2005.

[56] The PSD (2005). *A Draft National Strategy for the Sustainable Use of Plant Protection Products* (Part 2), http://www. pesticides. gov. uk/environment. asp?id=1524, accessed July 2005.

[57] *Control of Pesticides Regulations (COPR) 1986*, 4, 5, (a), Statutory Instrument SI 1986 No. 1510. See also *Plant Protection Products (Basic Conditions) Regulations (PPPR) 1997*, 4, 5(a), Statutory Instrument SI 1987 No. 189.

[58] *COPR 1986*, 5, b (i).

[59] *COPR 1986*, Schedule 3 (1). See also *PPPR 1997*, Schedule 3 (2) and COPR 1997, Schedule 3 (2).

[60] *COPR 1986*, Schedule 3 (4). See also *PPPR 1997*, Schedule 3 (3).

[61] *COPR 1986*, Schedule 3 (5). See also *PPPR1997*, Schedule 3 (3).

[62] *COPR 1986*, Schedule 3 (8). See also *PPPR 1997*, Schedule 1 and COPR 1997 Schedule 3 (1).

[63] *COPR 1997* Schedule 3 (4). See also *PPPR 1997*, Schedule 3 (4).

[64] *Control of Substances Hazardous to Health (COSHH) Regulations 2002*, Statutory instrument SI 2002 No. 2677.

[65] Ministry of Agriculture, Fisheries and Food (MAFF)/Health and Safety Commission (HSC) (1990), *The code of practice for the safe use of pesticides on farms and holdings (Green Code)*, HMSO.

[66] Central Science Laboratory (CSL) (2005), *A Survey of Current Farm Sprayer Practices in the United Kingdom 2004*, http://www. voluntaryinitiative. org. uk/_Attachments/Second%20Farm%20Application%20Practice%20Survey%20June%202005. pdf, accessed July 2005.

[67] Crop Protection Association (CPA). *Before Spraying:Spray Operator Training*, http://www. voluntaryinitiative. org. uk/_attachments/BPG%20pesticide%20Training. pdf, accessed July 2005.

[68] MAFF/HSC (1990), *The code of practice for the safe use of pesticides on farms and holdings (Green Code)*, HMSO.

[69] Ibid.

[70] Ibid.

[71] *COPR 1997*, Schedule 3 (4). See also *PPPR 1997*, Schedule 3 (4).

[72] *COPR 1986*, Schedule 3 (1). See also *PPPR 1997*, Schedule 3 (2) and *COPR 1997*, Schedule 3 (2).

[73] MAFF/HSC (1990), *The code of practice for the safe use of pesticides on farms and holdings (Green Code)*, HMSO.

[74] Ibid.

[75] Ibid.

[76] Evidence from the Royal College of General Practitioners, December 2004.

[77] Tenth Report, paragraph 2. 47.

[78] MAFF/HSC (1990), *The code of practice for the safe use of pesticides on farms and holdings (Green Code)*, HMSO.

[79] Defra, *Draft Code of Practice for the Safe Use of Plant Protection Products*, http://www. pesticides. gov. uk/main. asp?id=1046, accessed July 2005.

[80] Evidence from the HSE, October 2004.

[81] HSE, *Pesticides Incidents Report, 1 April 2003 – 31 March 2004*, http://www. hse. gov. uk/fod/pir0304. pdf, accessed April 2005.

[82] Evidence from the HSE, October 2004.

[83] HSE, *Pesticides Incidents Report, 1 April 2003 – 31 March 2004*, http://www. hse. gov. uk/fod/pir0304. pdf, accessed April 2005.

[84] Evidence from Robert Hadway, HSE, October 2004.

[85] Evidence from the Health and Safety Executive, October 2004.

[86] Agricultural Engineers Association (AEA), NSTS Achieves Target, http://www. aea. uk. com/sprayer/index. htm, accessed July 2005.

[87] Ibid.

[88] National Register of Spray Operators, http://nroso. nptc. org. uk accessed July 2005.

[89] The Voluntary Initiative, http://www. voluntaryinitiative. org. uk accessed July 2005.

[90] Environment, Food and Rural Affairs Committee, Eighth Report:Progress on the use of pesticides:the Voluntary Initiative, http://www. publications. parliament. uk/pa/cm200405/cmselect/cmenvfru/258/25802. htm accessed July 2005.

[91] Defra (2005), Set-aside Handbook and Guidance for England, http://www. defra. gov. uk/farm/capreform/pubs/pdf/Setaside-hb. pdf accessed July 2005.

[92] Ibid.

[93] Evidence from Agra CEAS, February 2005.

[94] Defra, *Environmental Stewardship*, http://www. defra. gov. uk/erdp/schemes/es/default. htm accessed July 2005.

[95] Defra, *Environmental Stewardship:look after your land and be rewarded*, http://www.defra.gov.uk/erdp/pdfs/es/es-promotional-booklet. pdf accessed July 2005.

[96] Pesticides Forum, *Pesticides Forum Paper PF 141:National Pesticides Strategy*, http://www. pesticides. gov. uk/pesticides_forum. asp?id=617, accessed April 2005.

[97] CSL (2005), *A Survey of Current Farm Sprayer Practices in the United Kingdom2004*, http://www. voluntaryinitiative. org.uk /_Attachments / Second%20Farm%20Application%20Practice%20Survey%20June%202 005. pdf accessed July 2005.

[98] Linking Environment and Farming (LEAF), *The LEAF Aims*, http://www. leafuk. org/leaf/organisation/default. asp?id=4030470 accessed April 2005.

[99] Linking Environment and Farming (LEAF) (2003), *The New LEAF Audit:A whole Farm business and environmental management tool*, http://www. leafuk. org/leaf/producers/audit. asp

[100] Linking Environment and Farming (LEAF), *Measuring Change: Environmental Improvement on farm over 10 years*, http://www. assuredproduce. co. uk/_code/common/item. asp?id=4032604 accessed May 2005.

[101] Beratungsgesellschaft für integrierte Problemlösungen (Bipro) (2004), *Assessing the Economic Impacts of the Specific measures to be part of the Thematic Strategy on the Sustainable Use of Pesticides* (ENV. C. 4/ETU/2003/0094R), European Commission, Brussels.

In: Crop Spraying and the Health of Residents...
Editor: The Royal Commission, pp. 111-119
ISBN: 1-60021-391-X

Chapter 6

CONCLUSIONS AND RECOMMENDATIONS

6.1. In this report we have looked at two areas:

- the scientific evidence on which the Department for Environment, Food and Rural Affairs (Defra) has based its decisions on bystander exposure and its policy on access to information on crop spraying; and
- the wider issues related to the handling and communication of uncertainty, and to public involvement, values and perceptions.

SCIENTIFIC EVIDENCE

6.2. The scientific evidence we considered covered four main areas:

Health

(a) the establishment of what are deemed to be safe levels of human exposure to specific pesticides;

(b) the assessment of reports of ill health amongst residents and bystanders which they consider to be related to the use of pesticides in agriculture;

Exposure

(c) the assessment of exposure on the basis of a standard approach to calculate the amount of pesticide a resident or bystander might be exposed to; and

Risk

(d) risk assessment to determine on the basis of the calculation in (c) whether in normal use residents and bystanders are exposed to pesticide levels within the limits identified at (a).

6.3. We have identified grounds for concern in respect of all of these areas.

HEALTH ISSUES

6.4. There is no dispute that some people who have been exposed to pesticides have become ill. The dispute has concerned the causality and underlying basis for these illnesses. On the evidence that we have received we cannot draw firm conclusions on causality. But we are persuaded that it is possible that some cases of ill health could, on further investigation, be shown to be due to complex effects following exposure to pesticides.

6.5. We are concerned that the toxicological testing currently undertaken within the pesticides approval and assessment process, whilst taking into account a wide range of health problems, does not encompass the full range of conditions that have been described to us by members of the public and attributed by them to exposure to pesticides. This situation is unsatisfactory; in effect we lack input data on 'no effect levels'for the ill health conditions not covered by the testing. In the absence of this data no scientific model of exposure can address those particular health concerns.

6.6. We recognise that it will be extremely difficult to develop testing arrangements that address these concerns, but believe that with better investigation of reported ill health it should be possible to devise improved test protocols. These, combined with better monitoring of reported ill health in patients, should lead to improved protection of human health.

6.7. We have been unimpressed by the arrangements for monitoring human health. At one level we find that there is little in the way of UK-based epidemiological studies that look at possible associations between pesticides and serious diseases such as cancer and degenerative neurological disease. We accept that such studies would be challenging but have been assured that they could be designed. They would clearly be made easier if information on the individual pesticides used were available. Our recommendations at (6. 20-6. 24) reflect our view that the reported health effects need to be taken more seriously.

6.8. Residents and bystanders have told us that their experiences of the current reporting arrangements under the Pesticide Incident Appraisal Panel (PIAP) of the Health and Safety Executive are unsatisfactory. Some have also reported difficulty in having their concerns taken seriously and accessing the most appropriate medical experts. Our understanding from talking to PIAP is that in effect it only monitors acute exposures; even here PIAP accepts that it understates the number of cases. We conclude that this system should be radically reformed, extended to cover chronic cases and integrated into the Health ProtectionAgency (HPA).

6.9. We have made recommendations to address our concerns on the monitoring and treatment of those who believe their health has been affected by pesticide exposure (6. 25-6. 28). The recommendations are important in themselves but are also intended to result in improved understanding of the underlying medical issues. Some of those who contacted us were exposed to pesticides from non-agricultural use, primarily wood treatment and amenity spraying. They have reported similar health problems and similar difficulties in obtaining treatment from the NHS. We have therefore recommended that these changes to the system (6. 20-6. 28) should also be extended to these patients (6. 29).

Exposure Assessment

6.10. We have significant concerns about the assessment of resident and bystander exposure.

Although the assessment is conservative in terms of the way in which toxicological data are reflected in exposure limits, it has been drawn up to represent a single situation and looks at one exposure route which it assumes to be the main route. It is deterministic, not taking account of a range of possible outcomes. More sophisticated probabilistic models are needed to allow risk managers to assess the likelihood that serious situations could arise and to manage exposure risks accordingly. Such a model should be drawn up as a matter of urgency and validated by extensive laboratory and field testing.

6.11. The current exposure assessment approach has not been rigorously evaluated under field conditions and has largely been assessed in relation to non-peer-reviewed experiments conducted on a limited scale over 20 years ago and reassessed on the basis of data collected for different purposes in Germany and the US. This is not satisfactory.

It is also important that field testing should take account of variation in spraying conditions, such as boom height, wind speed and tractor speed, so as to provide a sensitivity analysis and to enable legislators and operators to have a better understanding of which variables are important and how significant an impact they can make. Recommendations (6. 30-6. 32) advocate a more sophisticated approach underpinned by further research.

Risk Assessment

6.12. The risk assessment depends crucially on both the exposure assessment and toxicological testing. Although uncertainty factors are built in, they are there to cover issues related to the toxicology testing and cannot compensate for the weakness of the exposure assessment.

6.13. In chapter 2 we explain that the susceptibility of the population to a pesticide varies between individuals. In chapter 3 we explain that the level of exposure of a resident or bystander will also vary as a result of the combination of factors at the time of spraying and afterwards. Ill health in a proportion of those in areas where agricultural pesticides are used may well result from a person who is particularly sensitive being exposed to a high enough level of material to cause an effect. In effect, the distribution of susceptibility thresholds would overlap with the distribution of actual exposure levels.

We do not have the scientific data to demonstrate that this has happened, but it cannot be ruled out and we believe that action is necessary to reduce the risk of overlap and consequent ill health. As we cannot see a way of identifying and removing from potential exposure those who are inherently more susceptible, the only option is to reduce overall levels of exposure through the package of risk management measures recommended at (6. 33-6. 43) and through the spraying notification measures in 6. 51 and 6. 53.

GOVERNANCE ISSUES

6.14. In the light of the lack of rigour in the underlying science, we have been surprised at the level of confidence expressed in advice to Ministers and the level of assurance given to the public about the safety of residents and bystanders potentially exposed to agricultural pesticides. We have concluded that the level of these assurances is not robustly founded in scientific evidence. Limitations in the data and alternative views of the science, as well as political and ethical judgements implicit in this advice, all need to be clearly acknowledged.

6.15. We have concluded that, in line with best practice in Defra, delivery of policy on pesticides should be separated from policy formulation. At the moment both fall to a Defra agency, the Pesticides Safety Directorate (PSD). Such a situation can lead to potential conflicts of interest, contributing to mistrust, which may be further compounded by the PSD being partially dependent on industry funding to meet its costs. Our recommendations address these issues (6. 44-6. 50).

PUBLIC INFORMATION

6.16. It is clear to us that there is a demand from those concerned about pesticides for better access to information. Defra proposed that information on what had been sprayed should be made available through a third party only. We rejected this approach as untenable over 20 years ago in our Tenth Report, *Tackling Pollution – Experience and Prospects*, and were disappointed to find it still being used to justify policy today.

6.17. We welcome the pilot project to evaluate a scheme where farmers tell their neighbours when they intend to spray. But we are concerned that any practical problems could ultimately be used as a pretext for revisiting the decision to require such prior notification.

We would regard any backtracking on this as unacceptable. Indeed, we conclude that the information provided should be extended to tell people what is going to be sprayed. This information should also be made available to a resident or bystander or a scientific investigator after the event. We have made recommendations accordingly (6. 51-6. 53).

ECONOMIC ISSUES

6.18. We commissioned an independent analysis by agricultural consultants on the main economic impacts of our recommendations on the agricultural industry, which suggested that these would be minor in aggregate. Nevertheless we do recommend measures that could further alleviate the impact on the industry by allowing buffer zones to be used to meet farmers' set-aside obligations under Common Agricultural Policy (CAP) subsidy schemes or to qualify in principle under other support measures (6. 54). Our analysis suggests that this is something within the discretion given to the UK government by the EU common rules on these policies.

RECOMMENDATIONS

6.19. Our recommendations encompass a wide range of measures and reflect the significant degree of uncertainty, which in our view underlies current understanding of resident and bystander safety and exposure. We have recommended a more precautionary approach pending more research. This will deliver a significant strengthening of risk management procedures until a more rigorous evaluation of the risk and necessary accompanying measures can be made. The interim measures need to be introduced in the UK and drawn to the attention of the authorities in the EU who, we are told, do not currently regard bystander exposure to pesticides as a priority.

Health Issues

6.20. Based on the conclusions from our visits and our understanding of the biological mechanisms with which pesticides interact, it is plausible that there could be a link between resident and bystander pesticide exposure and chronic ill health. We find that we are not able to rule out this possibility. We recommend that a more precautionary approach is taken with passive exposure to pesticides. The existing uncertainties indicate an urgent need for research to investigate the size and nature of the problem and any underlying mechanisms that link pesticide spraying to ill health (2. 65).

6.21. We recommend a comprehensive systematic review of the literature on pesticide spraying and human health that takes account of the shortcomings of the Ontario Report (2. 66).

6.22. We recommend that an imaginative systematic approach is taken to apply both well validated as well as novel clinical investigative methods to those with chronic symptoms linked to pesticide spraying such as magnetic resonance spectroscopy (MRS) and gene and protein profiling (2. 67).

6.23. We recommend that the Health Protection Agency (HPA) and related organisations within the devolved administrations in Scotland and Wales collect population data on pesticides, their metabolites and biomarkers of effects that would provide a sound basis for exposure assessment and also could be used to establish a national database for monitoring (2. 68).

6.24. We recommend that the private sector and universities be encouraged to develop new animal models that better reflect the chronic disorders experienced by residents and bystanders exposed to pesticide spraying (2. 69).

Monitoring Human Health

6.25. We endorse the Department of Health's move to strengthen higher professional development in the field of toxicology within general practice. They should also ensure that professionals working in public health and specialised poisons centres have a clear awareness and understanding of how to investigate the chronic health problems related to pesticides by residents and bystanders (2. 97).

6.26. We recommend that the Royal Medical Colleges agree how patients with chronic symptoms associated with pesticide spraying should be investigated and treated, and identify a clear referral pathway from primary care to an appropriately trained consultant specialist (2. 98).

6.27. We recommend that a new national reporting and monitoring mechanism for ill health associated with pesticide spraying should replace the Pesticide Incident Appraisal Panel (PIAP), and that this should fall within the remit of the Health Protection Agency and related organisations in the devolved administrations (2. 99).

6.28. We recommend that newly acquired and standardised clinical, physiological and laboratory information from those who attribute their adverse health effect to passive pesticide exposure should form the basis for future biological monitoring. In addition to those exposed to agricultural spraying, we recommend the establishment of a system that places greater emphasis on surveillance for adverse health effects of pesticides (2. 100).

6.29. People are exposed to pesticides through uses other than in agriculture, such as amenity and timber treatment and these people have reported similar health concerns to those discussed in chapter 2. Although we have not studied these areas in detail, the measures we recommend on health issues and monitoring human health should be extended to cover these non-agricultural pesticides (2. 107).

Exposure Assessment

6.30. We recommend that the current approach for assessing resident and bystander exposure should with some urgency be replaced by a computational model which is probabilistic, looks at a wider range of possible exposure routes and more robustly reflects worst-case outcomes. The model should be rigorously validated by wind tunnel and field tests designed for the purpose, including non-standard conditions to test the sensitivity of the model predictions (3. 53).

6.31. We recommend that validation of the model should be supported by monitoring of representative pesticides in a range of field conditions. Such monitoring should include measurements of concentrations in air for extended periods after spraying to contribute to better understanding of possible exposure other than through immediate spray drift (3. 54).

6.32. We recommend that both the model and the test data be peer reviewed and published in the open scientific literature so that the basis for the predictions is transparent and can be evaluated (3. 55).

Reducing Exposure

6.33. We recommend that all actual spraying practice be brought into line with the aspirations of the Green Code recommendations (chapter 5), including giving proper regard to the importance of optimal timing of the application and therefore efficacy of the pesticide.

This will require appropriate monitoring arrangements, and sanctions for noncompliance (3. 56).

6.34. We recommend the introduction of buffer zones alongside residential property and other buildings such as schools, hospitals and retirement homes where people may be adversely affected by crop spray. Buffer zones should be evaluated on the basis of the revised resident and bystander exposure model. Until then, on the basis of our current understanding, we see five metre buffer zones as providing the necessary further protection and encourage the active management of these so as to decrease the likelihood of spray drift affecting residents and bystanders (5. 86).

6.35. We recommend that 'Grandfather rights', allowing those born before 31 December 1964 to spray without passing the relevant tests, should be phased out by 2010 (5. 68).

6.36. We recommend that the purchase of agricultural pesticides should be restricted to those who are permitted to use them (5. 69).

6.37. We recommend that the National Sprayer Testing Scheme becomes compulsory and that spray equipment should be checked annually so that it is regularly maintained in good working order by both contractors and farmers. Consideration should also be given to applying this recommendation to more rudimentary or small-scale types of spraying, such as those operating from all terrain vehicles (e. g. quad bikes) (5. 98).

6.38. We believe that adherence to some of the recommended conditions under the Green Code should become statutory duties. These include maximum wind speed, spraying practice as specified on the label, boom height and vehicle speed. Further research on refining the resident and bystander exposure model should lead to recommendations for revised spraying conditions for all factors relevant to minimising exposure, and thus to a revision of the statutory obligations (5. 73-5. 74).

6.39. We recommend that as part of their on-site visits the Health and Safety Executive Inspectors should request evidence of:record keeping of pesticide use; training and certification; and testing of spray equipment (5. 91).

6.40. We recommend that courts should be empowered to suspend or revoke the right to spray where operators have been convicted of breaking the law (5. 94).

6.41. We recommend that membership of the National Register of Spray Operators becomes compulsory and that it should be linked to possessing a Certificate of Competence, which should expire if continuing professional development is not undertaken (5. 99).

6.42. We recommend that the UK government press the European Commission to implement the recommendations on bystanders made by the former Scientific Committee on Plants, taking into account the concerns discussed in our report (5. 10).

6.43. The use of pesticides in organic farming is subject to the same regulations as in non-organic farming and our recommendations therefore apply equally to them (5. 112).

Governance Issues

6.44. We recommend that the advice to Ministers from technical and scientific advisory committees, such as that of the Advisory Committee on Pesticides on the approval of pesticides, should take account of the criteria employed regarding risk and precaution, the choices available, and the likely impacts on those affected by the decisions being recommended. This information needs to be reflected explicitly when communicating risk management decisions to the public. The meaning of such terms as 'adequate protection'should be clearly defined and communicated. Areas of scientific uncertainty, ignorance and indeterminacy need to be recognised and their nature described explicitly. It should be clear how uncertainty has been managed within the assessment, for example, through the use of probabilistic analysis (3. 53), uncertainty factors etc.

The effect of uncertainty on the estimate of risk should be clear (5. 37).

6.45. We recommend that advice passed to Ministers from expert committees and statements to the public should properly reflect the full range of opinion within such committees (5. 43).

6.46. We recommend that advice from committees should be differentiated between advice to the Minister, written in accessible language, and advice to departments, which would normally be for officials to handle (5. 28).

6.47. We recommend that in consultations and in dialogue with the public concerning pesticides, as in other areas of uncertainty, science should not be positioned without argument as being beyond doubt. It should be presented in a manner that is open about the level of risk and uncertainty involved (5. 47).

6.48. In accordance with government guidelines for best practice we recommend that consultation responses should be made available to the public, unless the respondent has said that they wish their response to remain private (5. 49).

6.49. We recommend that the draft National Pesticides Strategy should give full recognition to the need to safeguard human health and the environment (5. 56).

6.50. We recommend that responsibility for pesticides policy should be separate from that for the approval of pesticides. One way to do this might be to move the policy function from the Pesticides Safety Directorate to a unit within the Environment Directorate of Defra.

The unit should have a clear remit for recognising the importance of safeguarding human health and the environment as well as the requirements for pest control; the Department of Health should have an explicit role in agreeing its remit and assessing its performance. The remaining functions relating to the approval of pesticides could be transferred to the Environment Agency to enable better co-ordination with wider environmental objectives (5. 58).

Public Information

6.51. We recommend that records of which pesticides, and when and where they have been used, should be directly available from the persons responsible for crop spraying upon request to any resident and bystander and to researchers investigating the health effects of resident and bystander exposure (5. 84).

6.52. We recommend that the residents living next to fields that are to be sprayed be given prior notification of what substances are to be sprayed, where and when. The results of the pilot study in this area announced by the government should be treated as an exercise to determine how best to provide information, not as an opportunity to re-examine the principle of doing so, which should be accepted (5. 79).

6.53. We recommend that for those wishing to access land, notices should be provided on rights of way or at points where access is likely. These notices should specify what is sprayed and when. A simple sign saying land may be sprayed is not satisfactory, not least because it could be a disincentive to those wishing to access land at times when spraying may not have taken place for some time (5. 80).

Reducing the Cost

6.54. We recommend that the UK interpretation of the rules on the Single Farm Payment scheme should be changed to allow land in buffer zones to count towards set-aside and that farmers should be strongly encouraged to make use of agri-environment schemes to minimise the economic costs of buffer zones (5. 126).

In: Crop Spraying and the Health of Residents...
Editor: The Royal Commission, pp. 121-122
ISBN: 1-60021-391-X

APPENDIX A
WRITTEN MINISTERIAL STATEMENT

Alun Michael
The Minister for Rural Affairs and Local Environmental Quality
Department for Environment, Food and Rural Affairs
Date:16 June 2004

RISKS TO PEOPLE FROM PESTICIDE SPRAYING

I have now completed a detailed consideration of the outcome of two consultation exercises regarding the introduction of buffer zones and information on pesticide use. As a result I would like to announce a number of measures designed to provide additional reassurance about the pesticide regulatory process, and to improve public access to information about pesticides used.

Firstly, I would like to announce that the Royal Commission on Environmental Pollution has agreed to my request to carry out a special study to examine the scientific evidence on which Defra has based its decisions on the risks to people from pesticide exposure.

I requested this study following a detailed evaluation of the responses made to the two public consultations launched on 21 July 2003, which sought views on the introduction of buffer zones around residential properties near farms, and how residents can best be informed of the pesticides that are being sprayed.

My number one priority has always been to ensure that the safety arrangements we have in place for pesticides do the job of protecting the public. The independent scientific advice I have received, both from the independent Advisory Committee on Pesticides and Defra's Chief Scientific Adviser, is very clear that the existing system provides full reassurance. I am confident that no new scientific evidence has come to light that would provide a safety based justification for buffer zones, and I have therefore decided against their introduction around agricultural land.

Nevertheless the responses to the consultations demonstrate that there is a perception that the current arrangements, and in particular the assessment of risks, are inadequate. I have listened to the concerns of campaigners who hold strong views about how crop spraying has affected their health. Their views are undoubtedly sincerely held and although no new scientific evidence was produced to support their case, I believe the time is now right for a fresh and independent appraisal of the science. This is why I have asked the Commission to examine the scientific evidence on which the regulatory system is based. The Commission will be free to take a new and independent approach to the question. Further details about the scope and management of the Commission's study will be released in due course.

I have also decided to introduce new legal measures requiring farmers and growers to keep records of pesticides used on crops and to make those records available to the public via a third party. Officials are contacting general practitioners' and lawyers' representatives and those for Community Health Councils to explore whether they would act as third party representatives in such cases. These measures are at the early stage of development, and will require changes to existing pesticide regulations to put them into effect. I will bring forward proposals for parliamentary consideration at the earliest opportunity.

In parallel, a pilot study will be set up to explore how residents living next to farms can be notified in advance of pesticide use. The study will look at the practicality and cost of various options for how notification can take place. Officials are working on the terms of reference for the study and further details will be published as soon as practical. These measures are in addition to a biomonitoring study already being arranged to look at the effect of pesticides in the air. There is also an on-going study, co-ordinated by the Working Group on Risk Assessment of Mixtures of Pesticides (WIGRAMP) into the effect of numbers of pesticides used in combination.

I also intend to invite key stakeholders to a meeting in the near future to explain the reasoning behind the decisions made.

In: Crop Spraying and the Health of Residents...
Editor: The Royal Commission, pp. 123-124
ISBN: 1-60021-391-X

APPENDIX B ANNOUNCEMENT OF STUDY AND INVITATION TO SUBMIT EVIDENCE

ANNOUNCEMENT OF THE STUDY

On 16 June 2004, the Royal Commission on Environmental Pollution announced that it had agreed to a request from the Rt Hon Alun Michael, the then Minister of Rural Affairs and Local Environmental Quality, Department of Environment, Food and Rural Affairs, to carry out a special study on crop spraying and the health of residents and bystanders.

The Commission will examine the scientific evidence on which Defra has based its decision on bystander exposure and its policy on access to information on crop spraying. The Commission will also consider wider issues related to the handling and communication of risk and uncertainty, as well as public involvement, values and perceptions in this context.

On 3 August 2004, the Commission officially opened the consultation process and invited evidence on, and comments regarding, the intended scope of the study as well as recommendations for specific questions to be addressed. Written evidence submitted to the Commission will be made available to the public upon request, unless those submitting it specify otherwise.

INVITATION TO SUBMIT EVIDENCE

The new study will examine the scientific evidence on which Defra has based its decision on bystander exposure and its policy on access to information on crop spraying. The Commission will also consider wider issues related to the handling and communication of risk and uncertainty, as well as public involvement, values and perceptions in this context.

In accordance with the recommendations set out in the Commission's Twenty-first Report *Setting Environmental Standards*, the consultation aims to include people's values in both the framing of the problem and in the definition of the questions to be addressed. The Commission is therefore also inviting comments regarding the intended scope of the study and recommendations for specific questions to be addressed.

The Commission is requesting views and information that will help it to carry out the study. When the initial evidence has been considered, the Commission may request further written or oral evidence from relevant parties. The Commission has obtained the submissions made to the PSD consultation last year. If you submitted evidence to PSD and wish to refer to that submission as your response to this consultation exercise we would be happy to do so if you let us know.

Broad Topics to be Covered

In its study, the Commission intends to cover four braod topics:health dimention, scientific modelling, legal and policy issues and public concern.

Each topic covers a wide range of issues, some of which concern more than one topic. The issues the Commission aim to address are:

Health Dimension

What are the biological effects of bystander exposure to pesticides (what is the knowledge base)? What are the limits of toxicology and epidemiology in cases of bystander exposure to pesticides? How plausible is it that pesticides cause the health problems reported? What systems are in place to respond to and record bystander exposure and how well do they work (e. g. GP's, PIAP)?

Scientific Modelling

How robust is the pesticide bystander exposure model? What happens in the 'real' world (are rules followed, how good (old) is technology, how well does the model deal with dispersion)? How were safety factors derived? What kinds of exposure take place?

Legal and Policy Issues

How much nuisance should an individual have to bear? What are the potential legal issues surrounding harm, causation and compensation? Are there issues of pollution law etc? What role does Europe play in the policy process? What are the regulations regarding access to information and are they justified in the light of the cases of bystander exposure?

Public Concern

What are the public concerns regarding pesticide exposure, and what are the causes of concern? To what extent have the public been genuinely engaged in the bystander exposure policy process? What role do public values play? How are risk and uncertainty perceived and communicated? What is the relationship between science and the public in the bystander issue?

In: Crop Spraying and the Health of Residents…
Editor: The Royal Commission, pp. 125-130 ISBN: 1-60021-391-X

APPENDIX C
CONDUCT OF THE STUDY

In order to carry out this study the Royal Commission sought written and oral evidence, commissioned studies and advice on specific topics and made a number of visits.

EVIDENCE

In parallel with the news releases inviting evidence, which are reproduced in appendix B, the Secretariat wrote direct to a large number of individuals and organisations. The organisations and individuals listed below either submitted evidence or provided information on request for the purposes of the study or otherwise gave assistance. In some cases, indicated by an asterisk, meetings were held with Commission Members or the Secretariat so that oral evidence could be given or particular issues discussed.

GOVERNMENT DEPARTMENTS

Department of Environment, Food and Rural Affairs* Department of Health* Foreign and Commonwealth Office

Government Agencies

Health and Safety Executive* Health Protection Agency* Environment Agency Pesticides Safety Directorate*

Devolved Administrations

National Assembly for Wales

European and International Bodies

European Commission, Environment Directorate General European Commission, Research Directorate-General European Environment Agency

Other Associations

Advisory Committee on Pesticides*
Agra CEAS
Agricultural Engineers Association (AEA)*
Agricultural Industries Confederation (AIC)
Allergy UK
Assured Produce
Bayer CropScience Ltd
British Crop Production Association (BCPC)*
British Independent Fruit Growers'Association (BIFGA)
British Medical Association (BMA)
Cambridge Food Technologies
Central Science Laboratory (CSL)
Centre for Ecology and Hydrology (CEH)
Certis*
Chemical Industries Association (CIA)
Country Land and Business Association (CLA)
Crop Protection Association (CPA)*
DEMOS*
Economic and Social Research Council (ESRC)
Family Farmers Association
Farmline
Foxendown Fruit Farms
Friends of the Earth (FOE)
HillFairs Farming
Horticultural Development Council
Institute of Environment and Health (IEH)*
IPCS/OECD
J. Brock and Sons
J. G. Farms
Ketton Parish Council
Langmead Farms Ltd.
Lantra
LEAF
NationalAssociationofAgriculturalContractors(NAAC)*
National Farmers'Union (NFU)*
Met Office
Micron Sprayers Ltd*
Monsanto

PAN *UK**
Panztel UK ltd
H. R. Philpot and Son (Barleylands) Ltd
Processed Vegetable Growers'Association (PVGA)
Rambler'sAssociation
Robert Mitchell Partnership
Royal Society for the Protection of Birds
SAHSU*
Scottish Natural Heritage
SoilAssociation*
Syngenta*
Silsoe Research Institute*
Transport and General Workers'Union
Women's Food and Farming Union (WFU)
Worldwide Fund for Nature (WWF)

Individuals

AudreyAdcock
G. L. Anderson
MargaretAnderson
AnneArgent
GeoffArgent
KatherineArgent
Helen Beaumont
John Baker
Claus Berend
Daisy Bourne
Charlie Bransden
Frances N. Brown
Richard Bruce
Professor Jacquie Burgess*
Heather Cameron
Enfys M Chapman
Dr John Cherrie*
Miles Clarke
Caroline Clarke
Dr Charlie Clutterbuck*
David Coggon*
John Coyte
Professor Simon Deakin
Sir Liam Donaldson*
Suzanne Donaldson
Georgina Downs*
D. J. Ellis

Professor Samuel Epstein
Anne Evans
R. G. Field
Dr Marilyn Glenville
B. J. Gooderham
Dr Jules Griffin
Michael C. Groombridge
Keith and Jean Groombridge
W. A. Groombridge and Sons
Dr Edward Hamlyn
ProfessorTonyHardy
Dr Chris Hilson
M. F. and J. Holliday
Felicity Holmes
Professor Malcolm Hooper*
Terry Howard
Jacqueline Howard
Tom M. Howie
DrVyvyanHoward
ProfessorAlanIrwin
Diane Irwin
John Irwin
Brenda Jackson
Jim Jennings
Professor Kevin Jones*
Peter Kindersley
Peter King
Professor Brian Legg
Mr John Lowry
Professor Richard Macrory
Felicity Mankin
Brenda Marks
Professor Graham Matthews
ElizabethA. McDonagh
I. R. Mead
Professor Paul Miller*
Robert Montgomery
John Murphy
Francis Murphy
Dr Sarah Myhill
Annie Newitt
Ian Panton
Professor Judith Petts
Professor David Pimentel, Cornell University
Professor Colin Pritchard
Dr Iain Purchase

Brian Ralph
Andy Reynolds
A. Revell
Laurie Richie
Barbara Robinson
Rodney Rynell
Dr Margaret Sanborn
J. E. Saville
Dr Annette Scivier*
ProfessorAnthonySeaton
Richard Simkin
Mary Simmons
Peter H J Smith
Dr Christopher Stopes*
NealeThomas
KayWadey
Dr Gordon Walker*
Dr Hilary Walker
Professor Simon Wessely*
M. G. Whyles
Dr James Wilson*
JudyWilson
FrancesWolferstan
N. I. and P. G. Wright
Professor Michael Yost*

We received evidence from a small number of people who wish to remain anonymous.

We also received copies of the submissions of evidence to the 2003 PSD consultation on buffer zones and access to information.

COMMISSIONED STUDIES

The following studies were commissioned in the course of the study:

Report to the Royal Commission on Environmental Pollution on the Pesticides literature review (Ontario College of Family Physicians) - Dr Michael Burr. January 2005.

Potential economic and financial impact of buffer zones - Agra CEAS Consulting Ltd. March 2005.

Pesticides and Bystander Exposure:Tort Liability – J. Murphy. May 2005.

National estimates of residential land adjacent to arable land using Countryside Survey sample data – Centre for Ecology and Hydrology. April 2005

Visits

October 2004

Society of Chemical Industry Conference, *Making pesticides easier and more effective to use*. Farm visit, Wiltshire.

November 2004

Fifth Open Meeting of the Advisory Committee on Pesticides. Farm visit, Kent.

24 January – 1 February 2005

Visits to residents and bystanders homes to Devon, Gloucestershire, Oxfordshire, Lincolnshire, Norfolk, Essex, Hampshire, Isle of Wight.

January 2005

Visit to Silsoe Research Institute, followed by presentations and discussions. Farm visit, Norfolk.

Conference, *Science, Medicine and the Law*, London.

February 2005

Visit to residents and bystanders in Cambridgeshire. Farm visit, Cambridgeshire.

Environment, Food and Rural Affairs Committee, *Progress on pesticides evidence sessions*.

March 2005

Visit to Syngenta C. T. Laboratory, followed by presentations and discussions. Advisory Committee on Hazardous Substances open session.

In: Crop Spraying and the Health of Residents...
Editor: The Royal Commission, pp. 131-133
ISBN: 1-60021-391-X

APPENDIX D
PUBLIC MEETING

On 25 September 2004, the Commission hosted a seminar at Church House, Westminster, to gather views from interested parties relevant to deciding the scope of the study. The open meeting involved around 50 participants and the following programme, with an open discussion in the afternoon.

Introduction
Sir Tom Blundell, Chairman Royal Commission on Environmental Pollution
Pesticides and the public
Professor Alan Irwin, Dean of the Faculty of Social and Environmental Studies, University of Liverpool
Pesticide exposures for people in agricultural areas
Georgina Downs, Pesticides Campaigner
Why pesticides are used
Dr James Gilmour, Individual Member, British Crop Production Council (BCPC)
Pesticides regulation
Richard Davis, Director of Approvals, Pesticides Safety Directorate (PSD)
Scientific modelling for pesticides use
Dr Vyvyan Howard, Department of Human Anatomy and Cell Biology, University of Liverpool
Pesticide application and dispersion
Professor Paul Miller, Director of Technology, Silsoe Research Institute
Pesticides and health
Professor Stephen Holgate, MRC Clinical Professor of Immunopharmacology, University of Southampton and Member of the Royal Commission on Environmental Pollution
Open discussion
Contributions from the floor in response to the morning session and other issues of importance. *Chaired by Sir Tom Blundell*

In addition to the speakers and Members of the Commission, the other participants were:

Paul Andersen	Farmers Weekly
Brian Baxter	Member of the public
Graham Belson	National Association of Agricultural Contractors
Helen Bower	Womens'Farmers Union
David Buffin	Pesticide Action Network UK
Dr Clare Butler Ellis	Silsoe Research Institute
Caroline Clarke	The Burnhams Group
Alissa Cook	Soil Association Pesticides Policy Advisor
Alison Craig	Pesticide Action Network UK
Tony Fletcher	London School of Hygiene and Tropical Medicine
Cathy Garretty	Department for the Environment, Food and Rural Affairs
Dr Clive Gregory	Cardiff University
Jill Hewitt	National Association of Agricultural Contractors
Neil Kift	National Farmers Union
Francis Murphy	Member of the public
Madeleine Murphy	Member of the public
Ian Panton	Member of the public
Kate Perry	Department for the Environment, Food and Rural Affairs
Peter Sanguinetti	Crop Protection Association UK Ltd
Peter H. J. Smith	Mid Bucks Beekeepers Association
Dr Stuart Smith	Health and Safety Executive
Dr Christopher Stopes	Ecos Consultancy
Dr Roger Turner	Cambridge Food Technologies
Dr Corinne Vaughan	Food Standards Agency
Jake Vowles	Agricultural Engineers Association
Kay Wadey	Member of the public
Phil Wheeler	Member of the public
Hazel Doonan	Agricultural Industries Confederation
Prof. Peter Simpson	Parliamentary and Scientific Committee and Editor of Science in Parliament
Neal Thomas	Member of the public
Miles Clarke	Member of the public
Jean Muir	Member of the public
Julian Hargreaves	Farmer
Hugh Lowe	Farmer
Julian Randall	Agronomist
Chris Snary	Member of the public
Kevin Jones	Member of the public
Dr Sue Popple	Pesticides Safety Directorate
Paul Hamey	Pesticides Safety Directorate
Julian Hasler	National Farmers Union member
Jonathan Little	Environment, Food and Rural Affairs Committee Specialist
Dr Maurice de Billot	Crop protection Association
Peter Hingley	Certis Europe
Kathy Smyth	Member of the public
Professor Malcolm Hooper	University of Sunderland

C. Devrell	Member of the public
J. F. Hitchin	Action on Pesticides Exposure
J. Wheatley	Member of the public
Barbara Robinson	Member of the public and representative of Suffolk Youth and Parent Support Group

In: Crop Spraying and the Health of Residents...
Editor: The Royal Commission, pp. 135-138 ISBN: 1-60021-391-X

APPENDIX E
PESTICIDES: AN EXPLANATORY NOTE

E.1. Pesticides are widely used to increase the production of food, fibre and other agricultural and horticultural crops (e. g. biofuels, medicinal plants, flowers), to protect livestock and to facilitate other agricultural aims (e. g. managing conservation land). Non-agricultural uses include application to amenity land (e. g. parks, forests, public rights of way, pavements, golf courses), timber, and house and gardens. Pesticides can also be used in pet flea and tick shampoos and collars and in lice and fungal treatments for humans.

E.2. When the word 'pesticide'is used without modification, it primarily implies a material synthesised by humans. Synthetic pesticides are chemicals developed and manufactured for a specific property and use; many organisms also produce their own natural pesticides[1] and some inorganic substances that occur naturally are used as pesticides.[2]

E.3. Pesticides are usually classified according to the type of pest they are designed to control (e. g. fungicides, herbicides, insecticides, algicides, nematocides and molluscicides).

A pesticide product consists of two parts: active ingredients and co-formulants. Active ingredients are the chemicals that actually control the pest. Co-formulants are primarily solvents and carriers that help deliver the active ingredient to the target pest, and serve to enhance the utility of the product. Co-formulants may be liquids into which the active ingredient is dissolved, chemicals that keep the product from separating or settling, and compounds that help secure the pesticide to its target after application. It is not only the 'active ingredient'in the pesticide that may have a biological effect. For example, there is evidence that nonylphenol and octylphenol ethoxylates, introduced to the UK in the 1940s[1] and used as surfactants in pesticide formulations, have endocrine disrupting properties.[2] Although these chemicals will be phased out by the end of 2006[3] this raises questions about the safety of co-formulants. The combination of an active ingredient with a compatible co-formulant is referred to as a formulation. Pesticides are formulated for a number of different reasons. A pesticide active ingredient in a relatively pure form, ready for manufacturer's use, is rarely suitable for field application. An active ingredient must usually be formulated in a manner that increases pesticide effectiveness in the field, improves safety features or enhances handling qualities.[4]

---

[1] Derris, a large climbing plant that is mainly cultivated in the tropics for its roots,contains the natural insecticide rotenone.

Table E. 1. Class, mode of action and potential acute impact[6]

CLASS	EXAMPLE OF ACTIVE INGREDIENT[3]	MODE OF ACTION	METABOLISM AND EXCRETION	ACUTE UNINTENDED IMPACT IN HUMANS[7]
Herbicide	glyphosate	Non-selective systemic herbicide absorbed by foliage with rapid translocation throughout the plant. Inactivated on contact with soil. Acts on the shikimic acid pathway that is not found in animals. [8]	In mammals, following oral administration, very rapidly excreted unchanged; does not bioaccumulate.	A herbicide of low selectivity. Where poisoning in humans could be positively attributed to glyphosate, the symptoms recorded were stomach cramps, diarrhoea, vomiting (associated with erosion of the gastrointestinal tract) and, in the case of dermal exposure, skin irritation.
Fungicide	kresoxim-methyl	Broad spectrum fungicide with protective, curative, eradicative and long residual disease control; acts by inhibiting spore germination. Redistribution via the vapour phase contributes to activity. Inhibits mitochondrial respiration by blocking electron transfer.	In animals, orally administered compound was widely distributed and quickly eliminated; no bioaccumulation. The major routes of excretion were faeces and urine; 32 different metabolites were identified.	No data.
Rodenticide	warfarin	Anti-coagulant rodenticide. Repeated ingestion is necessary to produce toxic symptoms General internal bleeding is induced by reduction of the prothrombin content of the blood.	Only slightly dangerous to humans and domestic animals when used as directed.	Doses of 1-2 mg/kg bodyweight for periods of 6-15 days have caused serious illness and death in humans. Serious illness was induced by the ingestion of 1. 7 mg/kg bodyweight, /day for six consecutive days with suicidal intent. All signs and symptoms were caused by haemorrhage.
Soil-sterilant	Methyl bromide	Multi-purpose fumigant used for insecticidal, acaricidal, and rodenticidal control. The mode of action of methyl bromide is still not understood. Proposed mechanisms of toxicity include the direct cytotoxic effect of the intact methyl bromide molecule or toxicity due to one of its metabolites. [9]	Metabolism not totally elucidated; inorganic bromide ion is formed.	A neurotoxic and narcotic agent with a characteristic delayed action. Fatal poisoning in humans have resulted from exposures to relatively high concentrations (from 33, 000 mg/m^3upwards) of methyl bromide vapour. Mild and moderate inhalation exposure may cause headache, dizziness, abdominal pain, nausea, vomiting, chest pain and difficulty in breathing.

[2] Copper sulphate,often sold as 'Bordeaux'or 'Burgundy'mixture,is used as a fungicide despite its toxicity to humans and other organisms.

[3] Active ingredient.

Table E. 1. Class, mode of action and potential acute impact[6] (Continued)

CLASS	EXAMPLE OF ACTIVE INGREDIENT[4]	MODE OF ACTION	METABOLISM AND EXCRETION	ACUTE UNINTENDED IMPACT IN HUMANS[7]
Insecticide	chlorpyrifos	Non-systemic organophosphate insecticide with contact, stomach and respiratory action; A cholinesterase inhibitor.	In rats, dogs and other mammals following oral administration, rapid metabolism occurs. Excretion is principally in the urine.	Reduction of plasma cholinesterase activity.
	carbaryl	Carbamate insecticide with contact and stomach action and slight systemic properties; weak/reversible cholinesterase inhibitor.	In mammals does not accumulate in body tissues, but is rapidly metabolised to non-toxic substances. Eliminated predominantly in the urine and faeces.	Blurred vision, headache, stomach ache and vomiting have been reported among workers not using proper precautions. Clinical symptoms in humans may include increased bronchial secretion, excessive sweating, salivation and lacrimation, pinpoint pupils, bronchoconstriction, abdominal cramps, vomiting, diarrhoea, bradycardia, muscle fasciculation, diaphragm and respiratory muscles paralysis, tachycardia, headache, dizziness, anxiety, mental confusion, convulsion, coma, and depression of the respiratory centre. Signs of intoxication develop quickly after absorption.
	cyfluthrin	Non-systemic pyrethroid insecticide with contact and stomach action. Acts on the nervous system of insects, and disturbs the function of neurons by interaction with the sodium channel.	In animals largely and very quickly eliminated; 97% of the administered amount was eliminated after 48 hours via the urine and the faeces.	The symptoms and signs of acute poisoning resulting from exposure to different pyrethroids are similar. Clinical analysis of cases of acute pyrethroid poisoning due to occupational or accidental exposure revealed symptoms including burning, itching and tingling sensations of the skin, which resolved after several hours.

E.4. In general, active ingredients are less volatile than their co-formulants. When a distinctive smell is reported on property close to sprayed fields, it is usually due to these more volatile co-formulants and does not mean that significant amounts of the active ingredients will also be found in that area. It is usually thought that active ingredients are more likely to move onto adjacent land through spray drift than in volatile form.

[4] Active ingredient.

There are some exceptions where the active ingredient is more volatile but most of these active ingredients are no longer permitted for use.

E.5. Pesticides have one or more standard or common name(s) (e. g. chlorpyrifos) and one or more chemical name(s) (e. g. 0, 0-diethyl 0-(3, 5, 6-trichloro-2-pyridyl) phosphorothioate), and different companies make products using the same active ingredients but with registered trade names (e. g. Dursban). Pesticides can be broadly classified according to chemical type; for example the agricultural insecticides can be described as organophosphates, carbamates, organochlorines and pyrethroids.[5]

E.6. Table E. 1 shows examples of different pesticide classes, and their mode of action on the pest organism and non-target animals. The examples of pesticide active ingredients have been primarily taken from agricultural use, although they may also have wider uses, for example in the home and garden.

REFERENCES

[1] Pirie, D. , Steven, L. , McGrory, S. , Best, G. *Survey of Hormone Disrupting Chemicals.* SEPA. August 1996.

[2] http://www. defra. gov. uk/news/2003/031015c. htm accessed May 2005.

[3] http://www. pesticides. gov. uk/environment. asp?id=1434 accessed May 2005.

[4] Martin, A. , Whitford, F. , Jordan, T. , (2001). *Pesticides and formulation technology.* Ed. Blessing, A. Purdue Pesticide Programs.

[5] http://www. epa. gov/pesticides/about/types. htm accessed February 2005.

[6] British Crop Production Council (BCPC). *A world compendium:The e-pesticide manual* 13th Ed. Editor C S Tomlin.

[7] WHO/FAO data sheets on pesticides- http://www. inchem. org/

[8] Stenersen, J. (2004). *Chemical pesticides:mode of action and toxicology.* CRC Press.

[9] WHO/FAO data sheets on pesticides- http://www. inchem. org/

In: Crop Spraying and the Health of Residents...
Editor: The Royal Commission, pp. 139-140
ISBN: 1-60021-391-X

APPENDIX F
CHRONIC FATIGUE SYNDROME

F.1. Chronic fatigue syndrome (CFS) or myalgic encephalomyelitis (ME) is a complex and multifactorial disorder characterised by severe debilitating fatigue, sleep abnormalities, impaired memory and concentration, temperature disturbance and musculoskeletal pain.[1] In the Western world the frequency has been estimated to be as high as about 0.5% in the general population[2] and 0. 07% in children.[3] Although now recognised as a clinical entity, a significant research effort has failed to identify any single set of abnormalities that can account for the symptoms.

F.2. Although there is no known aetiology, nor a single diagnostic marker, a large number of diverse factors has been proposed as contributing to pathogenesis, including persistent virus infection, immune activation with lymphocyte abnormalities, exposure to toxins, pesticides and other chemicals, stress, and neuroendocrine dysfunction.[4]

F.3. An increasing number of laboratory and physiological abnormalities identified in CFS is strengthening the case for the disorder having, at least in part, an organic basis.

Application of messenger RNA transcript array technology to identify gene expression in circulating cells of CFS cases has revealed a number of specific genes that are either over- or under-expressed, indicative of a genetic and molecular 'fingerprint'for CFS.[5,6] Other abnormalities recently reported include increased neutrophil programmed cell death, elevations in circulating levels of transforming growth factor _, [7] reduced microvascular responses to acetylcholine,[8] reduction in serotonin (5-hydroxytrypta-mine a major neurotransmitter) transporters,[9] characteristic urinary electrophoretic profiles[10] and spinal fluid abnormalities.[11]

F.4. In the light of these more recent findings, it is recognised that the 1994 US Centers for Disease Control (CDC) definition of CFS could be improved upon to identify more homogeneous disease sub-groups within the CFS population that would be of value for further clinical and laboratory research and in directing patient management.[12]

REFERENCES

[1] A report of the CFS/ME Working Group. January 2002. Department of Health. (http://www. doh. gov. uk/cmo/cfsmereport/).

[2] Fukuda, K. , Straus, SE. , Hickie, I. , Sharpe, M. C. , Dobbins, J. G. , Komaroff, A. (1994). The chronic fatigue syndrome: a comprehensive approach to its definition and study. *Annals of Internal Medicine.* 121 (12) 953-9.

[3] Dowsett, E. G. , and Colby, J. (1997). Long-term sickness disease due to ME/CFS in UK schools. *Journal of Chronic Fatigue Syndrome*, 3:29-42.

[4] Komaroff, A. L. , and Buchwald, D. (1998). Chronic fatigue syndrome:an update. *Annual Reviews of Medicine.* 49:1-13.

[5] Powell, R. , Ren, J. , Lewith, G. , Barclay, W. , Holgate, S. , and Almond, J. (2003). Identification of novel expressed sequences, upregulated in the leucocytes of Chronic Fatigue Syndrome Patients. *Clinical and Experimental Allergy* 33:1450-6.

[6] Whistler, T, . Unger, E. R. , Nisenbaum, R. , and Vernon, S. D. (2003). Integration of gene expression, clinical and epidemiologic data to characterise Chronic Fatigue Syndrome. *Journal of Translational Medicine* 1:10-18.

[7] Kennedy, G. , Spence, V. , Underwood, C. , and Belch, J. J. F. (2004). Increased neutrophil apoptosis in chronic fatigue syndrome. *Journal of Clinical Pathology* 57:891-3.

[8] Spence, V. A. , Khan, F. , Kennedy, G. , Abbot, N. C. , and Belch, J. J. F. (2004). Acetylcholine mediated vasodilatation in the microcirculation of patients with chronic fatigue syndrome. *Prostaglandins, Leukotrienes and Essential Fatty Acids.* 70:403-407.

[9] Yamamoto, S. , Ouchi, Y. , Onoe, H. , Yoshikawa, E. , Tsukada, H. , Takahashi, H. , Iwase, M. , Yamaguti, K. , Kuratsune, H. , and Watanabe, Y. (2004). Reduction of serotonin transporters of patients with chronic fatigue syndrome. *Neuroreport.* 15 (17):2571-4.

[10] Casado, B. , Zanone, C. , Annovazzi, L. , Iadarola, P. , Whalen, G. , and Baraniuk, JN. (2005). Urinary electrophoret-ic profiles from chronic fatigue syndrome and chronic fatigue syndrome/ fibromyalgia patients:a pilot study for achieving their normalization. *Chromatography B - Analytical Technologies in the Biomedical and Life Sciences* 814 (1):43-51.

[11] Natelson, B. H. , Weaver, S. A. , Tseng, CL. , and Ottenweller, J. E. (2005) Spinal fluid abnormalites in patients with chronic fatigue syndrome. *Clinical and Diagnostic Laboratory Immunology.* 12(1) 52-5.

[12] Kennedy, G. , Abbot, N. C. , Spence, V. , Underwood, C. , and Belch, J. J. (2004) The specificity of the CDC-1994 Criterion for Chronic Fatigue Syndrome: Comparison of health status in those groups with patients who fulfil the criteria. *Annals of Epidemiology* 14: 95-100.

In: Crop Spraying and the Health of Residents…
Editor: The Royal Commission, pp. 141-143
ISBN: 1-60021-391-X

APPENDIX G MULTIPLE CHEMICAL SENSITIVITY

G.1. Multiple chemical sensitivity (MCS) also known as environmental chemical intolerance, is a disorder with multiple recurrent symptoms associated with exposure to chemicals (usually organic) at low concentrations that are otherwise well tolerated by the majority of people.[1] These chemicals include, alcohol, drugs, caffeine, food additives, carpet and furniture odours, fuel odours and engine exhaust, painting materials, perfume and scented products, and pesticides.[2] Its population prevalence, in the US, has been estimated to be as high as 12-15%[3] and it is more common in females.[4]

G.2. While the medical profession is divided over its existence as a clinical syndrome.[5] In 1999 COT reviewed the available evidence on MCS, but concluded that it was not possible to define the condition.[6] While, some believe it is a psychological condition, research has identified neurophysiological mechanisms that go some way to explain the symptoms. Recent studies have revealed increased sensitivity of mucous membranes,[7] altered odour perception,[8,9] Pavlovian conditioning,[10] and increased autonomic arousal in response to odour and chemical mixtures.[11] Toxicant-induced loss of tolerance (TILT)[12] relating to individual differences in susceptibility to neurohormonal sensitisation (amplification) triggered by chemical exposure may also contribute.

G.3. It is suggested that chemical, biological and psychological stimuli initiate sensitisation, while subsequent activation of the limbic and mesolimbic pathways in the brain facilitate altered behavioural, autonomic and immunological functions.[13] The influence of higher brain functions is also considered important, including negative thoughts and fear regarding a specific odour.[14,15] and a high frequency of psychiatric disorders.[16,17] However, in the US prevalence study[18] it is worthy of note that emotional problems prior to exposure was 1. 4% compared to 37. 7% after. Positron Emission Tomography (PET) scans of the brain in patients with MCS are reported to be normal,[19] although there are no reports of magnetic resonance spectroscopy (MRS) studies such as those carried out in GulfWar Syndrome patients (appendix I).

G.4. The recent description of a discriminant inhalation test using capsaicin (extract of hot pepper)[20] provides some evidence that enhanced neurogenic inflammation plays a role in MCS.[21] A genetic predisposition for MCS involving altered biotransformation of environmental chemicals[22] also suggests that there is still much to find out about this complex disorder.

REFERENCES

[1] Bolt, H. M. , and Kiesswetter, E. (2002) Is multiple chemical sensitivity a clinically defined entity? *Toxicology Letters*. 128 (1-3):99-106.

[2] Beers, M. and Berkow, R. (1999). *The Merck Manual of Diagnosis and Therapy*. Seventeenth Edition.

[3] Caress, S. M. and Steinemann, A. C. (2004). Prevalence of multiple chemical sensitivities:a population-based study in the southeastern United States. *American Journal of Public Health*. 94 (5):746-7.

[4] Caress, S. M. , Steinemann, A. C. , and Waddick, C. (2002). Symptomatology and etiology of multiple chemical sensitivities in the southeastern United States. *Archives of Environmental Health*. 57 (5):429-36.

[5] ltenkirch, H. (2000). Multiple chemical sensitivity (MCS) differential diagnosis in clinical neurotoxicology:a German perspective. *Neurotoxicology*. 21 (4):589-97.

[6] Department of Health (1999). *Annual report of the Committees on Toxicity, Mutagenicity, Carcinogenicity of Chemicals in Food, Consumer Products and the Environment*. p. 30.

[7] Osterberg, K. , Orbaek, P. , Karlson, B. , Akesson, B. , and Bergendorf, U. (2003). Annoyance and performance during the experimental chemical challenge of subjects with multiple chemical sensitivity. *Scandinavian Journal of Work, Environment and Health*. 29 (1):40-50.

[8] Ojima, M. , Tonori, H. , Sato, T. , Sakabe, K. , Miyata, M. , Ishikawa, S. , and Aizawa, Y. (2002) Odour perception in patients with multiple chemical sensitivity. *Tohoku Journal of Experimental Medicine*. 198 (3):163-73.

[9] Greene, GJ. , and Kipen, H. M. (2002). The vomeronasal organ and chemical sensitivity:a hypothesis. *Environmental Health Perspectives*. 110 Supplement 4:655-61.

[10] Otto, T. , and Giardino, ND. (2001). Pavlovian conditioning of emotional responses to olfactory and contextual stimuli:a potential model for the development and expression of chemical intolerance. *Annals of the New York Academy of Sciences*. 933:291-309.

[11] Fiedler, N. , and Kipen, H. M. (2001). Controlled exposures to volatile organic compounds in sensitive groups. *Annals of the New York Academy of Sciences*. 933:24-37.

[12] Miller, C. S. (2001). Toxicant-induced loss of tolerance. *Addiction*. 96 (1):115-37.

[13] Bell, I. R. , Baldwin, C. M. , and Schwartz, G. E. (1998). Illness from low levels of environmental chemicals: relevance to chronic fatigue syndrome and fibromyalgia. *American Journal of Medicine*. 105 (3A):74S-82S.

[14] Devriese, S. , Winters, W, . Van Diest, I. , De Peuter, S. , Vos, G. , Van de Woestijne, K. , and Van den Bergh, O. (2004). Perceived relation between odors and a negative event determines learning of symptoms in response to chemicals. *International Archives of Occupational and Environmental Health* 77 (3):200-4.

[15] Staudenmayer, H. , Binkley, K. E. Leznoff, A. , and Phillips. S. (2003). Idiopathic environmental intolerance:Part 2:A causation analysis applying Bradford Hill's criteria to the psychogenic theory. *Toxicological Reviews*. 22 (4):247-61.

[16] Bornschein, S. , Hausteine, C. , Zilker, T, . and Forstl, H. (2002). Psychiatric and somatic disorders and multiple chemical sensitivity (MCS) in 264 'environmental patients'. *Psychological Medicine*. 32 (8):1387-94.

[17] Eberlein-Konig, B. , Przybilla, B. , Kuhnl, P. , Golling, G. , Gebefugi, I. , and Ring, J. (2002). Multiple chemical sensitivity (MCS) and others:allergological, environmental and psychological investigations in individuals with indoor air related complaints. *International Journal of Hygiene and Environmental Health* 205 (3):213-20.

[18] Caress, S. M. , and Steinemann, A. C. (2004). Prevalence of multiple chemical sensitivities:a population-based study in the southeastern United States. *American Journal of Public Health*. 94 (5):746-7.

[19] Bornschein, S. , and Hausteiner, C. (2002). PET in patients with clear-cut multiple chemical sensitivity (MCS). *Nuklearmedizin*. 41 (6):233-9.

[20] Johansson, A. , Lowhagen, O. , Millqvist, E. , and Bende, M. (2002). Capsaicin inhalation test for identification of sensory hyperreactivity. *Respiratory Medicine*. 96 (9):731-5.

[21] Kimata, H. (2004). Effect of exposure to volatile organic compounds on plasma levels of neuropeptides, nerve growth factor and histamine in patients with self-reported multiple chemical sensitivity. *International Journal of Hygiene and Environmental Health*. 207 (2):159-63.

[22] McKeown-Eyssen, G. , Baines, C. , and Cole, D. E. (2004). Case-control study of genotypes in multiple chemical sensitivity:CYP2D6, NAT1, NAT2, PON1, PON2 and MTHFR. *International Journal of Epidemiology*. 33 (5):971-8.

In: Crop Spraying and the Health of Residents...
Editor: The Royal Commission, pp. 145-146 ISBN: 1-60021-391-X

Appendix H
Applications of Magnetic Resonance Spectroscopy to Toxicology and the Clinical Assessment of the Impact of Pesticides[1]

H.1. Nuclear magnetic resonance spectroscopy (MRS) has found widespread use in both medical diagnostics and as a research tool for monitoring biological systems.

Its major virtue for studying metabolism *in vivo* is that it is a non-invasive analytical approach that is not biased to a particular class of metabolite. In terms of studying the potential neurological effects of pesticides the approach has been used in two different modes. Within the clinical environment MRS has been used to study metabolism *in vivo* in humans, especially in the brain.

H.2. Metabolism can be monitored *in vivo* by MRS using either hydrogen or phosphate atoms. Because of the relative insensitivity of the approach only high concentration metabolites are detectable using clinical situations. However, the technique has been particularly useful in monitoring brain metabolism because of its non-invasive manner, the lack of motion artefacts in this organ (compared to the heart), and the reasonably homogeneous nature of the brain. Furthermore, the metabolites observable are highly indicative of neurodegeneration. By combining these approaches with magnetic resonance imaging (MRI) (especially in the form of chemical shift imaging) it becomes possible to diagnose these disorders using MRS. While MRI would be used initially to detect a tumour, MRS has been used to characterise tumour tissue types (e. g. distinguishing different types of brain tumours) and monitoring tumour cell metabolism (e. g. during treatment to induce apoptosis).

H.3. Metabonomics (or metabolomics) defines a global profile of the metabolites in a cell, tissue or organism using either hydrogen MRS or mass spectrometry in conjunction with statistical pattern recognition. The concept is to measure all the small molecule concentrations through a global analytical approach, and then to apply pattern recognition techniques to define a metabolic phenotype or 'metabotype. ' MRS has proven to be highly diagnostic in determining the organ of effect after administration of toxic compounds, with different toxins giving different patterns in the MRS spectra. Along with toxicogenomics, COT has agreed that there is a potential for applying these new toxicological methods further.[2,3]

REFERENCES

[1] Evidence from Dr Jules Griffin, Cambridge University, March 2005.

[2] http://www. advisorybodies. doh. gov. uk/cotnonfood/toxicogenomics. htm accessed June 2005.

[3] Battershill, J. M. (2005). Toxicogenomics:regulatory perspective on current position. *Human and Experimental Toxicology,* 24 (1):35-40.

In: Crop Spraying and the Health of Residents...
Editor: The Royal Commission, pp. 147-150 ISBN: 1-60021-391-X

APPENDIX I
LESSONS FROM THE GULF WAR SYNDROME

I.1. Fourteen years after the end of Operation Desert Storm, a substantial proportion of veterans of the First Gulf War continue to experience chronic, often debilitating conditions characterised by persistent headache, cognitive problems, somatic pain, fatigue, and gastrointestinal (GI), respiratory and skin problems. Such ill health was classified for 13 years as a form of post-traumatic stress disorder. The Research Advisory Committee on Gulf War Veteran's Illnesses (RACGWVI) established by US

Congress in 1998 issued a report in 2004[1] recognising that GWS constituted a serious health problem that could not be explained adequately by deployment stress or psychological factors and that a growing body of evidence indicated that an important component was neurological in character. Included in evidence cited by the Committee was a two-fold risk in the development of motor neurone disease.[2]

I.2. Among the prominent risk factors which the Committee found to be associated with GWS was excess pesticide exposure. It also drew attention to previous reports of a 'cerebro-organic delayed effect syndrome' characterised by fatigue, headache, GI problems, chemical intolerances and memory deficits.[3,4] The report identified combinations of pesticides (e. g. DEET and permethrin) and pyridostigmine bromide (PB) (anti-nerve agent medication) that synergistically increased indicators of neurotoxicity and behavioural change.[5] Other studies indicated that absorption of some pesticides (e. g. DEET and permethrin) was increased by concurrent exposure to jet fuels.[6]

I.3. Brain imaging using magnetic resonance spectroscopy (MRS, appendix H) revealed that the ratio of N-acetyl aspartate to creatinine, a measure of functional brain mass, was significantly lower in the brain stem and right and left basal ganglia of GWS patients compared to matched veteran controls.[7,8] The specific regions of the brain affected differed between those with different symptom profiles (Haley profiles 1-3). MRS abnormalities have also been reported in the hippocampal region,[9] a part of the brain involved in spatial orientation, memory and mood. Other objective abnormalities included dysregulation of the autonomic nervous system, [10, 11] disordered vestibular function, [12] peripheral nerve dysfunction[13] (although this has recently been disputed), [14] slowed neurobehavioural performance[15] and evidence of cognitive dysfunction. [16]

I.4. Biochemical studies have shown that those with GWS had reduced activity of paraoxonase (PON1) an enzyme that inactivates acetylcholinesterase inhibitor (AChEi) compounds including some pesticides. [17, 18, 19] Relatively rare genetic variation in a second enzyme involved in pesticide metabolism – butylcholinesterase – has also been identified disproportionately in GWS subjects. [20, 21] Investigators have proposed a mechanism by which exposure to AChEi or acute psychological stress leads to overproduction of a normally rare 'R'variant of the enzymes associated with altered cognitive and behavioural function and fatigue. [22] Most recently GWS veterans possessing the DD variant of the angiotensin converting enzyme (ACE) have been shown to be at eight times greater risk of developing symptoms of CFS than those with the II genotype. [23]

I.5. Among its many recommendations, the Advisory Committee report emphasised the importance of further research underpinned by a strategic approach to collect information efficiently on those affected and set a national goal to develop treatments for chronic multisystem illness. While the report was ambitious in its content and recommendations and has met with some criticism, [24, 25, 26] it has helped open up the debate over the investigation and management of patients with chronic multisystem disorders associated with pesticides and other environmental exposures.

REFERENCES

[1] Research Advisory Committee on GulfWar Veterans'Illnesses. (2004). Scientific progress in understanding Gulf War Veterans'Illnesses:report and recommendations.http://www1.va.gov/rac-gwvi/docs/ Report and Recom-mendations_2004. pdf

[2] Haley, R. W. (2003). Excess incidence of ALS in young GulfWar veterans. *Neurology.* 61 (6):750-756

[3] Ismail, K. , Everitt, B. , Blatchley, N. , Hull, L. , Unwin, C. , David, A. , and Wessely, S. (1999). Is there a GulfWar Syndrome. *Lancet.* 353 (9148) 179-182.

[4] Steele, L. (2000). Prevalence and patterns of Gulf War illness in Kansas veterans:association of symptoms with characteristics of person, place, and time of military service. *American Journal of Epidemiology.* 152 (10) 922-1002.

[5] Abdel-Rahman, A. , Abou-Donia, S. , El Masry, E. , Shetty, A. , andAbou-Donia, M. (2004). Stress and combined exposure to low doses of pyridostigmine bromide, DEET, and permethrin produce neurochemical and neurochemical and neuropathological alternations in cerebral cortex, hippocampus and cerebellum. *Journal of Toxicology and Environmental Health Part A.* 67 (2):163-192.

[6] Riviere, J. , Monteiro-Riviere, N. , and Baynes, R. (2002). GulfWar related exposure factors influencing topical absorption of 14-C permethrin. *Toxicology Letters* 135 (102):61-71.

[7] Haley, R. W, . Marshall, W. W. , McDonald, G. G. , Daugherty, M. A. , Petty, F. , and Fleckenstein, J. L. (2000). Brain abnormalities in Gulf War syndrome:evaluation with 1H MR spectroscopy. *Radiology.* 215:807-817.

[8] Meyerhoff, D. J. , Lindgren, J. , Hardin, D. , Griffis, J. M. , and Weiner, M. W. (2001). Reduced N-acetylaspartate in the right basal ganglia of ill Gulf War veterans by magnetic resonance spectroscopy. *Proceedings of the International Society of Magnetic Resonance Medicine.* 9:994.

[9] Menon, P. M. , Nasrallah, H. A. , Reeves, R. R. , and Ali, J. A. (2004). Hippocampal dysfunction in GulfWar syndrome. A proton MR spectroscopy study. *Brain Research.* 1009:189-194.

[10] Haley, R. W, . Vongpatanasin, W. , Wolfe, G. I. , Bryan, W. W. , Armitage, R. , Hoffmann, R. F. , Petty, F. , Callahan, T. S. , Charuvastra, E. , Shell, W. E. , and Marshall, W. W, . Victor, R. G. (2004). Blunted circadian variation in autonomic regulation of sinus node function in veterans with Gulf War syndrome. *American Journal of Medicine.* 117 (7): 469-478.

[11] Peckerman, A. , LaManca, J. J. , Smith, S. L. , Taylor, A. , Tiersky, L. , Pollet, C. , Korn, L. R. , Hurwitz, B. E. , Ottenweller, J. E. , and Natelson, B. H. (2002). Cardiovascular stress responses and their relation to symptoms in Gulf War veterans with fatiguing illness. *Psychosomatic Medicine.* 62:509-516.

[12] Roland, P. S. , Haley, R. W. , Yellin, W. , Owens, K. , Shoup, A. G. (2000). Vestibular dysfunction in Gulf War syndrome. *Otolaryngology- Head and Neck Surgery.* 122:319-329.

[13] Jamal, G. A. , Hansen, S. , Apartopoulos, F. , and Peden, A. (1996). The "Gulf War syndrome". Is there evidence of dysfunction in the nervous system? *Journal of Neurology, Neurosurgery and Psychiatry.* 60:449-451.

[14] Joseph S. C. (1997). A comprehensive clinical evaluation of 20, 000 Persian Gulf War veterans. *Military Medicine.* 162:149-155.

[15] Anger, W. K. , Storzbach, D. , Binder, L. M. , Campbell, K. A. , Rohlman, D. S. , McCauley, L. , Kovera, C. A. , and Davis, K. L. (1999). Neurobehavioral deficits in Persian Gulf veterans:evidence from a population-based study. *Journal of the International Neuropsychological Society.* 5:203-212.

[16] Storzbach, D. , Campbell, K. A. , Binder, L. M. , McCauley, L. , Anger, W. K. , Rohlman, D. S. , and Kovera, C. A. (2000). Psychological differences between veterans with and without Gulf War unexplained symptoms. Portland Environmental Hazards Research Center. *Psychosomatic Medicine.* 62:726-735.

[17] Haley, R. W, . Billecke, S. , and La Du, B. N. (1999). Association of low PON1 type Q (type A) arylesterase activity with neurologic symptom complexes in Gulf War veterans. *Toxicology and Applied Pharmacology.* 157:227-233.

[18] Mackness, B. , Durrington, P. N. , and Mackness, M. I. (2000). Low paraoxonase in Persian Gulf War veterans self reporting Gulf War syndrome. *Biochemical and Biophysical Research Communications.* 276:729-733.

[19] Hotopf, M. , Mackness, M. I. , Nikolaou, V. , Collier, D. A. Curtis, C. , David, A. , Durrington, P, . Hull, L. , Ismail, K. , Peakman, M. , Unwin, C. , Wessely, S. , and Mackness, B. (2003). Paraoxonase in Persian Gulf War veterans. *Journal of Occupational and Environmental Medicine.* 45:668-675.

[20] Lockridge, O. and Masson, P. (2000). Pesticides and susceptible populations:people with butyrylcholinesterase genetic variants may be at risk. *Neurotoxicology.* 21:113-126.

[21] Lockridge, O. (1999). Project Summary:Butyrylcholinesterase genetic variants in persons with Gulf War illness. Available at:http://www. gulflink. osd. mil/medsearch/GeneticStudies/DoD60. shtml.

[22] Cohen, O. , Erb, C. , Ginzberg, D. , Pollak, Y. , Seidman, S. , Shoham, S. , Yirmiya, R. , and Soreq, H. (2002). Neuronal overexpression of "read through"acetylcholinesterase is

associated with antisense-suppressible behavioral impairments. *Molecular Psychiatry.* 7:874-885.

[23] Vladutiu, G. and Natelson, B. (2004). Association of medically unexplained fatigue with Ach insertion/ deletion polymorphism in Gulf War veterans. *Muscle and Nerve*. 30 (10):38-43

[24] Deahl, M. (2005). Smoke, mirrors and Gulf War illness. *Lancet.* 365 (9460):645-638.

[25] Dyer, O. (2004). US and UK scientist disagree about causes of Gulf War Syndrome. *British Medical Journal* 329 (7472):940.

[26] Couzin, J. (2004). Epidemiology:VA advisers link Gulf War illnesses to neurotoxins. *Science.* 306 (5693):26-27

In: Crop Spraying and the Health of Residents...
Editor: The Royal Commission, pp. 151-152
ISBN: 1-60021-391-X

APPENDIX J
GREEN CARD PILOT SCHEME

J.1. The Health and Safety Executive (HSE) commissioned a pilot study to assess the frequency and severity of acute pesticide poisoning in the UK. This was undertaken by the National Poisons Information Service (NPIS) in Birmingham between 1990 and 1993. The driving force was the 1987 House of Commons Agriculture Select Committee report *Effects of Pesticides on Human Health*, which recommended that data on suspected cases of pesticide poisoning should be collated from pre-existing sources and evaluated.

J.2. The study also involved the introduction of a two-year pilot monitoring scheme (Green Card Scheme) in the West Midlands and Trent regions, using freepost 'green cards', or a dedicated pesticide telephone hotline, to be used by general practitioners and accident and emergency clinicians to record pesticide incidents. [1] HSE concluded that the information generated by the study was not sufficiently substantial to warrant development of the pilot into a permanent national system.

J.3. A parallel Yellow Card Scheme is currently used for post-marketing surveillance of new drugs. Another form of pharmacovigilance is Prescription Event Monitoring (PEM), a non-interventional observational cohort technique that prompts all primary care medical practitioners using a new drug to report any adverse event following their use by collecting details of prescriptions for the first 20, 000-50, 000 patients given the new drug, and sending a questionnaire to general practitioners after a suitable time interval.[2]

J.4. The Yellow Card system for pharmacovigilance differed from the pesticide pilot-monitoring scheme in the West Midlands, which was not statutory and relied on general practitioners independently reporting suspected adverse effects of pesticides.[3] There were a number of weaknesses in this system. Firstly, the scheme was not a nationwide programme but focused on the West Midlands and Trent regions which are not the most heavily agricultural areas in the UK. Secondly, the system was passive and relied on the general practitioner recognising that the cause of any ill health effects, that might be chronic and ill defined, were indeed due to pesticide exposure and actively reporting this to NPIS (Birmingham). The key incentive for the general practitioner would have been access to immediate specialist advice via a dedicated hotline, but without accurate information on which pesticide the patient had been exposed to even this would not have been useful.

REFERENCES

[1] Pesticide Monitoring Unit, National Poisons Information Service (Birmingham Centre). (1993). Surveillance of human acute poisoning from pesticides 1st October 1990-30th September 1993. HSE Commissioned Research. (the Vale Report).

[2] Prescription-event monitoring, drug safety research unit. http://www. dsru. org/pem2002. html accessed February 2005.

[3] Pesticide Monitoring Unit, National Poisons Information Service (Birmingham Centre) (1993).

In: Crop Spraying and the Health of Residents...
Editor: The Royal Commission, pp. 153-155
ISBN: 1-60021-391-X

APPENDIX K PESTICIDE DATA PACKAGES[1]

K.1. The main components of the data package typically required for a new pesticide to be approved for use in the UK fall into the following seven areas:

PHYSICO-CHEMICAL PROPERTIES

K.2. The applicant requesting pesticide approval is required to specify the chemical composition of the product, its active ingredient and any significant impurities that it may contain. Information must also be supplied on the physico-chemical properties of the active ingredient, and on the methods by which it can be detected and measured, for example in foodstuffs and water.

POTENTIAL TOXICITY IN HUMANS

K.3. Data on potential toxicity are required for the active ingredient, the product as a whole, and also any important metabolites of the active ingredient to which humans might be exposed. These data are derived largely from tests on laboratory animals, but care is taken to ensure that the use of laboratory animals is at the minimum level strictly necessary. If reliable information can be obtained by other means, it is used in preference.

DIETARY INTAKE

K.4. One of the ways by which a pesticide might cause harm to humans is through its presence as a residue in food. A particular concern is the potential for residues in food derived directly from any crops to which pesticides are applied or that might arise in the meat of animals that have been fed on a treated crop. In assessing the risk from pesticide residues in foods, it is necessary to identify and take account of all foodstuffs in which significant residues might occur, including those resulting from the use of other products that contain the same active ingredient. Additionally, if the use of a pesticide produces significant concentrations of toxic metabolites in food (i. e. substances formed by its chemical degradation in plants or animals), the acceptability of exposure to each of these metabolites is also assessed.

Exposures to Operators, Other Workers And Bystanders

K.5. The other circumstance in which human exposure to pesticides commonly occurs is in the course of their application or through contact with crops or other materials that have been treated with them. Estimating the profile of exposure in operators, other workers and bystanders is complex, and takes into account many factors including the physical form of the pesticide (e. g. liquid or granules) and the way (including frequency and duration) it is used.

Environmental Fate and Behaviour

K.6. In order to assess the potential impact of a pesticide on the environment, it is necessary to establish what happens to it once it has been applied; that is where it gets to, how fast it is degraded and by what mechanisms, and whether any of its degradation products might occur at levels sufficient to pose a risk. In particular, information is needed about the concentrations of the pesticide and any relevant breakdown products that will occur in soil, water and air, and the persistence of such contamination. Predicted environmental concentrations are derived and are used to assess the exposure of nontarget species in soil and water; the possible contamination of drinking-water supplies; the possible contamination of groundwater; and the potential for effects on, or residues in, following crops.

Ecotoxicology

K.7. The other major determinant of a pesticide's environmental impact is its toxicity to wildlife. The environmental risk assessment focuses upon possible effects of the pesticide on various populations of non-target organisms including:birds, wild mammals, fish, aquatic invertebrates and plants, insects (including bees) and other arthropods, earthworms and soil micro-organisms.

K.8. The risk to non-target organisms is generally assessed by the use of standard laboratory tests that give an indication of the toxicity of a particular active ingredient or formulation. These are used to derive the (the doses/concentrations at LD_{50}/LC_{50} which 50% mortality occurs in an acute toxicity study) and no observed effect level (NOEL)/no observed effect concentrations (NOEC) (the levels or the concentrations at which no effect is observed in longer-term studies).

Efficacy and Risk to Following Crops

K.9. Consideration of product efficacy is an integral part of the risk assessment process. If there are no discernible benefits from the application of the pesticide there is no reason to use the pesticides and thus no acceptable level of exposure. Data must be available to demonstrate the efficacy of the pesticide against target organisms when it is used in accordance with the label instructions. In addition, the application of pesticides (especially herbicides) to a crop may pose a risk to the crop itself or to adjacent or following crops. This risk is assessed through the evaluation of both laboratory and field trial data.

References

[1] Pesticides Forum, *Pesticides Forum paper pf 141:National Pesticides Strategy*, http://www. pesticides. gov. uk/pesticides_forum. asp?id=617 accessed July 2005.

In: Crop Spraying and the Health of Residents... ISBN: 1-60021-391-X
Editor: The Royal Commission, pp. 157-165

GLOSSARY

Absorption

Penetration of a surface layer, such as the skin (dermal absorption), leading to subsequent uptake into the body of the organism.

Acceptable Operator Exposure Level (AOEL)

The maximum amount of active ingredient to which the operator may be exposed without any adverse health effects.

Acetylcholine(ACh)

A key chemical in neurons (nerve cells) that acts as a neurotransmitter and carries information across the synaptic cleft (the space between two nerve cells).

Acetylcholinesterase(AChE)

An enzyme that breaks down the neurotransmitter acetylcholine at the synaptic cleft (the space between two nerve cells) so the next nerve impulse can be transmitted across the synaptic gap. Pesticides of the organophosphate and carbamate types act to paralyse and kill insects by inhibiting their acetylcholinesterase.

Active ingredient

A chemical that has been developed for its general or specific action against harmful organisms or on plants, parts of plants or plant products. Active ingredients are usually formulated with other chemicals in a pesticide product.

Adsorption

The taking up of one chemical onto the surface of another substance or particle.

Advisory Committee on Pesticides (ACP)

The independent statutory body that advises UK government Ministers on all matters relating to the control of pests.

Aerial application

The application of a pesticide from an aircraft (fixed-wing or helicopter) in flight.

Angioedema
An accumulation of fluid under the skin, usually due to an allergic response.

Annex I Directive 91/414/EEC
A positive list of active ingredients considered by Member States and the European Commission, and accepted as meeting criteria as specified in Council Directive 91/414/EEC.

Biocide
Any substance or micro-organism, including a virus or a fungus, having general or specific action on or against harmful organisms, e. g. mould killers. Biocides are regulated by the Health and Safety Executive.

Biomarker
A biomarker is a biological or biochemical response to an environmental chemical which gives a measure of exposure and sometimes also of a toxic effect.

Biological monitoring
Also known as biomonitoring, this is a scientific technique for assessing human exposure to natural and synthetic chemicals based on sampling and analysis of an individual's tissues and fluids.

Buffer zone
Also known as a no-spray zone or buffer strip, this is a strip of land adjoining a field boundary or environmental feature which does not receive chemical inputs.

Bystander
Any person who is present in (or near) the area where a pesticide is being (or has been) applied who is not directly involved in using the pesticide.

Carbamates
Esters of carbamic acids used as insecticides.

Case-control study
A scientific study that compares a group of people with disease (such as leukaemia) to a similar group of people without that disease. This type of study compares the levels of exposure (radiation or chemical) for each group before appearance of the disease.

Cholinesterase
Enzyme that breaks down acetylcholine after it is released at the synapse. It is inhibited by organophosphates. Cholinesterase can refer to one of two enzymes, acetylcholinesterase and pseudocholinesterase. (See acetylcholinesterase).

Cholinesterase inhibitor
Also known as anticholinesterase, a substance that inhibits a cholinesterase enzyme and thus prevents transmission of nerve pulses from one nerve cell to another or to a muscle.

Cluster

A cluster is the occurrence of a greater than expected number of cases of a disease such as cancer within a small area.

Co-formulant

A chemical within a pesticide formulation, other than the active ingredient.

Cognition

Mental functions, such as the ability to think, reason, and remember.

Cohort

A well-defined group of people who have had a common experience or exposure, who are then followed up for the incidence of new diseases or events, as in a cohort or prospective study.

Confounding

An inability to distinguish the separate impacts of two or more individual variables on a single outcome.

Control of Pesticides Regulations (COPR) 1986 (as amended)

The Regulations prohibit the sale, supply, storage, advertisement or use of pesticides unless Ministers in the responsible departments approve them, and unless general obligations (set out in Schedules to the Regulations) and conditions specific to individual pesticides (conditions of approval) are met.

Deterministic model

A model in which a single measure of risk is predicted from using fixed input data, e. g. toxicity and exposure data.

Directive 91/414/EEC

The Council Directive governing the marketing and authorisation of plant protection products in the Member States of the EU.

Endocrine disrupter

A substance that adversely affects hormonal processes.

Entrainment

The transport of one substance, e. g. droplets of pesticide, in another, e. g. vapour.

Epidemiology

The statistical study of categories of persons and the patterns of diseases from which they suffer, with the aim of determining the events or circumstances causing these diseases.

European Food Safety Authority (EFSA)

The European Food Safety Authority provides independent scientific advice on all matters linked to food and feed safety, including animal health and welfare and plant protection, and provides scientific advice on nutrition in relation to European Community legislation.

Exposure

Exposure to a chemical is the contact of that chemical with the outer boundary of the human body, i. e. the skin, openings into the body such as the mouth and nostrils, and punctures and lesions in the skin.

Exposure assessment

Procedure to establish quantitatively the extent of exposure in any particular situation.

Food and Environment Protection Act (FEPA) 1985

The legal arrangements for the evaluation and approval of pesticides are laid down in the Food and Environment Protection Act 1985 (as amended). Section 16 (Part III) of the Act gives Ministers broad powers to protect the health of human beings, creatures and plants; safeguard the environment; to secure safe, efficient and humane methods of controlling pests; and make information about pesticides available to the public. These are the UK national rules governing pesticides.

Food Standards Agency (FSA)

The independent food safety watchdog set up by an Act of Parliament in 2000 and implemented by the Food Standards Act 1999 to protect the public's health and consumer interests in relation to food.

Formulated product

Active pesticides ingredient plus other 'inert'co-formulants.

Fungicide

A pesticide used to control unwanted fungal diseases.

Genotoxin

A chemical or other agent that damages cellular DNA resulting in mutations or cancer.

Green Code

Common shorthand for the *Code of Practice for the Safe Use of Pesticides on Farms and Holdings*. The Green Code provides advice to assist farmers, growers and other professional users of pesticides to meet their legal obligations and other standards of good practice. The Green Code is authorised by statute but its provisions are not directly legally binding.

Gulf War Syndrome (GWS)

A widely-used term referring to unexplained illnesses occurring in Gulf War veterans.

Hazard

The set of inherent properties of a substance, or mixture of substances, that makes it capable of causing adverse effects to humans, other organisms or the environment.

Health and Safety Executive (HSE)

The government organisation responsible for the evaluation of non-agricultural pesticides, rodenticides and biocidal products.

Herbicide

A pesticide used to control unwanted vegetation; also known as weed killer.

In vitro

Biological techniques or processes performed outside a living organism.

In vivo

Biological techniques or processes performed within a living organism.

Intra-uterine

Within the uterus (womb).

Insecticide

A pesticide used to control unwanted insects.

Latency

Time from the first exposure of a chemical until the appearance of a toxic effect.

Leukaemia

Form of cancer resulting in an overproduction of abnormal white blood cells that suppresses normal red blood cell and platelet production.

Local Environment Risk Assessment for Pesticides (LERAP)

The calculation required before you can reduce the aquatic buffer zone requirement for specific pesticide products using horizontal boom or broadcast air-assisted sprayers.

Longitudinal study

A study that involves the repeated observation or examination of a set of subjects over time.

Lymphoma

Cancer of cells of the lymphatic system, a network of lymph vessels, lymph nodes, tonsils and the spleen involved in fighting infection.

Meta-analysis

A quantitative method of combining the results of independent studies (usually drawn from the published literature) and synthesising summaries and conclusions.

Microphthalmia

Rare disorder, usually inherited, in which one or both eyes are abnormally small.

Modelling

An analytical technique used to simulate processes to explain the impact of one or more factors on a number of outcomes; includes both deterministic and probabilistic modelling.

National Proficiency Tests Council(NPTC)

The Council provides a full range of UK vocational qualifications for the land-based and related industries.

National Sprayer Testing Scheme (NSTS)

A sprayer testing scheme to meet various requirements, including assurance schemes, processor/retailer production protocols, as well as to satisfy the desire of many operators to verify the accuracy and overall fitness of their pesticide application requirement. This has been extended under the industry-led Voluntary Initiative.

Neurasthenia

A condition characterised by chronic fatigue and weakness, loss of memory, and generalised aches and pains, formerly thought to result from exhaustion of the nervous system.

Neuropathy

A problem in peripheral nerve function (any part of the nervous system except the brain and spinal cord) that causes pain, numbness, tingling, swelling and muscle weakness in various parts of the body.

Neurotoxic

Capable of harming nerve tissue.

No observed adverse effect level (NOAEL)

The highest exposure level in a toxicity study at which there are no statistically significant and/or biologically significant increases in the frequency of adverse effects between the group of animals exposed to the test substance and its respective control group.

National Register of Sprayer Operators (NRoSO)

A central register of sprayer operators using continuing professional development as a means of ensuring ongoing training.

Oncogenic

Causing or giving rise to tumours.

Operator

The person mixing, loading and applying the pesticide.

Organochlorine

Chlorine-containing organic chemical used for pest control (including insecticides, herbicides, fungicides).

Organophosphate (OP)

Term is usually used in pest control to describe derivatives of phosphoric and similar acids. However, it may be more accurate to describe such compounds as 'organophosphorus compounds'. This term covers a wide range of chemicals not all of which display insecticidal or anticholinesterase activity.

Pathophysiology

The study of how normal physiological processes are affected by disease.

Persistent organic pollutants (POPs)

A class of chemicals that persist in the environment, are capable of long-range transport, bioaccumulate in human and animal tissue, and impact human health and the environment.

Personal protective equipment (PPE)

Any device or appliance designed to be worn or held by an individual for protection against one or more health and safety hazards.

Pesticide

Defined as 'any substance, preparation or organism prepared or used, among other uses, to protect plants or wood or other plant products from harmful organisms; to regulate the growth of plants; to give protection against harmful creatures; or to render such creatures harmless'(Food and Environment Protection Act 1985). They include biocides, insecticides, herbicides, rodenticides, fungicides, termaticides and repellents. Some pesticides can be highly specific while others are designed to be effective against a wide spectrum of pests.

Pesticides Safety Directorate (PSD)

The executive agency of the Department for Environment, Food and Rural Affairs responsible for evaluation and regulation of agricultural pesticides.

Plant protection product

An active substance or preparation containing one or more active substances, formulated as it is supplied to the user, intended to:

- protect plants or plant products against harmful organisms or prevent the action of such organisms;
- influence the life processes of plants other than as a nutrient (e. g. as a growth regulator);
- preserve plant products, in so far as such substances or products are not subject to the provisions of EU law on preservatives;
- destroy unwanted plants;
- destroy parts of plants or check or prevent the undesired growth of plants.
- Sometimes used as a synonym for 'pesticide', but not in the strict legal sense.

Plant Protection Product Regulations (PPPR) 2003

The Regulations that implement Council Directive 91/414/EEC in the UK.

Probabilistic model

A model which incorporates variability and uncertainty by using input data with probability distributions rather than fixed values to output an estimated distribution for the measure of concern.

Pyrethroid

Synthetic insecticide structurally related to the natural pyrethrin insecticides found in pyrethrum flowers.

Rapporteur

The Member State that is responsible for the evaluation of active substances prior to consideration at the EU Commission Standing Committee on Food Chain and Animal Health.

Risk

The possibility that a harmful event arising from exposure to a substance, or mixture of substances, may occur under specific conditions.

Risk assessment

The process of decision making applied to problems where there are a number of possible outcomes and it is uncertain which outcome will happen.

Rodenticide

A pesticide use to control mice and other rodents.

Sorption

A general term used for the processes of absorption and adsorption.

Spray drift

The movement of a pesticide through the air at the time of application or soon thereafter, to any site other than that intended for application (often referred to as off-target).

Standing Committee on the Food Chain and Animal Health (SCFA)

The European Commission-led body on which all Member States are represented, which votes on whether or not pesticide active ingredients be included in Annex I of Directive 91/414/EEC.

Surface tension

A property possessed by liquid surfaces whereby they appear to be covered by a thin elastic membrane in a state of tension. The surface tension of a sprayed liquid pesticide can be affected by co-formulants.

Synergistic

An effect of two chemicals acting together which is greater than the simple sum of their effects when acting alone:such chemicals are said to show synergism.

Systemic

Used to describe a chemical that affects the whole organism. A systemic farm dip, for example, might penetrate through the skin and reach the central nervous system.

Tort

Damage, injury, or a wrongful act done wilfully, negligently, or in circumstances involving strict liability, but not involving breach of contract, for which a civil suit can be brought.

Toxicity

The quality of being poisonous, especially the degree of adverse effects. The severity of toxicity produced by any chemical is related to the exposure concentration and the exposure time. This relationship varies with the developmental stage of an organism and with its physiological status.

- **Acute toxicity:** adverse effects occurring within a short time of administration of a single dose of a chemical, or immediately following short or continuous exposure, or of multiple doses over 48 hours or less.
- **Chronic toxicity:** adverse effects occurring as a result of repeated daily dosing of a chemical, or exposure to the chemical, for part of an organism's lifespan. With experimental animals, this usually means a period of exposure of more than three months.

Toxicology

The scientific study of poisons.

Toxicogenomics

The application of knowledge of genes associated with disease states to the study of the toxicology of chemical and physical agents.

Vapour pressure (vpa)

The pressure exerted by a vapour in equilibrium with its solid or liquid phase.

In: Crop Spraying and the Health of Residents...
Editor: The Royal Commission, pp. 167-168
ISBN: 1-60021-391-X

ACRONYMS

ACP	Advisory Committee on Pesticides
ALSPAC	Avon Longitudinal Study of Parents and Children
AOEL	acceptable operator exposure level
BIPRO	Beratungsgesellschaft für integrierte Problemlösungen
BMA	British Medical Association
CCDC	Consultant in Communicable Disease Control
CFS	chronic fatigue syndrome
ChE	cholinesterase
COC	Committee on Carcinogenicity of Chemicals in Food, Consumer Products and the Environment (Department of Health)
COPR	TheControl of Pesticides Regulations 1986 (as amended)
COSHH	TheControl of Substances Hazardous to Health Regulations 2002 (as amended)
CPA	Crop Protection Association
CPD	continuing professional development
Defra	Department for Environment, Food and Rural Affairs
DH	Department of Health
DWP	Department for Work and Pensions
EC	European Community
EFSA	European Food Safety Authority
EMAS	Employment Medical Advisory Service (Health and Safety Executive)
EPA	Environmental Protection Agency (US)
ES	Environmental Stewardship
EU	European Union
FEPA	The Food and Environment Protection Act 1985 (as amended)
FSA	Food Standards Agency
GAP	good agricultural practice
GIS	geographic information system
GLP	Good Laboratory Practice
GLPMA	Good Laboratory Practice Monitoring Authority
GP	general practitioner
GWS	GulfWarSyndrome
HPA	Health Protection Agency
HPS	Health Protection Scotland
HSE	Health and Safety Executive
LEAF	Linking Environment and Farming
LERAP	Local Environment Risk Assessment for Pesticides

LOEL	lowest observed effect level
MAFF	Ministry of Agriculture, Fisheries and Food
MCS	multiple chemical sensitivity
ME	myalgic encephalomyelitis
MRI	magnetic resonance imaging
MRS	magnetic resonance spectroscopy
NHS	National Health Service
NMR	nuclear magnetic resonance
NPTC	National Proficiency Tests Council
NPIS	National Poisons Information Service
NOAEL	no observed adverse effect level
NOEL	no observed effect level
NRoSO	National Register of Sprayer Operators
NSTS	National Sprayer Testing Scheme
OECD	Organisation for Economic Co-operation and Development
PAN UK	Pesticide Action Network UK
PIAP	Pesticide Incidents Appraisal Panel
POP	persistent organic pollutant
PPPD	Plant Protection Products Directive (EU)
PPE	personal protective equipment
PPR	Panel on Plant Health, Plant Protection Products and their Residues (EU)
PSD	Pesticides Safety Directorate
SAHSU	Small Area Health Statistics Unit
SANCO	Santé et Protection des Consommateurs (Health and Consumer Protection Directorate-General of the European Commission)
SCPH	Standing Committee on Plant Health (EU)
SCFA	Standing Committee on the Food Chain and Animal Health (EU)
SCP	Scientific Committee on Plants (EU)
SSTS	Scottish Skills Testing Service

In: Crop Spraying and the Health of Residents...
Editor: The Royal Commission, pp. 169-170
ISBN: 1-60021-391-X

MEMBERS OF THE COMMISSION

Sir Tom Blundell (Chair)

Chair of School of Biological Sciences and Sir William Dunn Professor and Head of Department of Biochemistry, University of Cambridge and Professorial Fellow, Sidney Sussex College.

Professor Roland Clift

Distinguished Professor of Environmental Technology and Director, Centre for Environmental Strategy, University of Surrey.

Professor Paul Ekins

Head, Environment Group, Policy Studies Institute. Professor of Sustainable Development, University of Westminster.

Sir Brian Follett

Chair, Teacher Training Agency; Chair, Arts and Humanities Research Board; Professor, Department of Zoology, University of Oxford.

Dr Ian Graham-Bryce

Principal and Vice Chancellor Emeritus, University of Dundee.

Professor Stephen Holgate

Medical Research Council Clinical Professor of Immunopharmacology, University of Southampton.

Professor Brian Hoskins

Royal Society Research Professor and Professor of Meteorology, University of Reading.

Professor Jeffrey Jowell QC

Professor of Public Law, University College London.

Professor Susan Owens

Professor of Environment and Policy, University of Cambridge, Department of Geography, and Fellow of Newnham College.

Professor Jane Plant

Professor of Geochemistry, Imperial College of Science, Technology and Medicine.

Professor Steve Rayner

Director, James Martin Institute for Science and Civilization, Professor of Science and Civilization, Saïd Business School, University of Oxford and Professorial Fellow, Keble College.

Mr John Speirs

Chairman, Chemistry Leadership Council's Futures Group.

Professor Janet Sprent

Emeritus Professor of Plant Biology, University of Dundee Board Member, Scottish Natural Heritage

Secretariat

Secretary to the Commission

Tom Eddy

Assistant Secretaries

Jo Bray

Charlotte Grey Dan Hamza Goodacre Philippa Powell

In: Crop Spraying and the Health of Residents...
Editor: The Royal Commission, pp. 171-172
ISBN: 1-60021-391-X

PREVIOUS REPORTS BY THE ROYAL COMMISSION ON ENVIRONMENTAL POLLUTION

25th report	Turning the Tide - Addressing the Impact of Fisheries on the Marine Environment	Cm 6392, December 2004
Special report Biomass as a Renewable Energy Source		2004
24th report	Chemicals in Products – Safeguarding the Environment and Human Health	Cm 5827, June 2003
Special report The Environmental Effects of Civil Aircraft in Flight		2002
23rd report	Environmental Planning	Cm 5459, March 2002
22nd report	Energy – the Changing Climate	Cm 4749, June 2000
21st report	Setting Environmental Standards	Cm 4053, October 1998
20th report	Transport and the Environment – Developments since 1994	Cm 3752, September 1997
19th report	Sustainable Use of Soil	Cm 3165, February 1996
18th report	Transport and the Environment	Cm 2674, October 1994
17th report	Incineration of Waste	Cm 2181, May 1993
16th report	Freshwater Quality	Cm 1966, June 1992
15th report	Emissions from Heavy Duty Diesel Vehicles	Cm 1631, September 1991
14th report	GENHAZ – A system for the critical appraisal of proposals to release genetically modified organisms into the environment	Cm 1555, June 1991
13th report	The Release of Genetically Engineered Organisms	Cm 720, July 1989
12th report	Best Practicable Environmental Option	Cm 310, February 1998
11th report	Managing Waste:The Duty of Care	Cm 9675, December 1985
10th report	Tackling Pollution – Experience and Prospects	Cm 9149, February 1984
9th report	Lead in the Environment	Cm 8852, April 1983
8th report	Oil Pollution of the Sea	Cmnd 8358, October 1981
7th report	Agriculture and Pollution	Cm 7644, September 1979
6th report	Nuclear Power and the Environment	Cm 6618, September 1976
5th report	Air pollution Control:An Integrated Approach	Cm 6371, January 1976
4th report	Pollution Control:Progress and Problems	Cm 5780, December 1974

3rd report	Pollution in some British Estuaries and Coastal Waters	Cm 5054, September 1972
2nd report	Three Issues in Industrial Pollution	Cm 4894, March 1972
	First Report	Cm 4585, February 1971

More information can be obtained from http://www. rcep. org. uk or from the Secretariat at Third Floor, 5-8 The Sanctuary, Westminster, London SW1P 3JS.

The Royal Commission on Environmental Pollution is an independent body, appointed by the Queen and funded by the government, which publishes in-depth reports on what it identifies as the crucial environmental issues facing the UK and the world.

Information on the Commission's work can be found at www. rcep. org. uk. Copies of this report are available to download from the website, alternatively, the Commission can be contacted at: Royal Commission on Environmental Pollution 5 - 8 The Sanctuary Westminster London SW1P 3JS

Email:enquiries@rcep. org. uk

INDEX

A

B

C

D

E

F

G

H

I

J

K

L

M

N

O

P

T

U

V

W

Y